Keepin

CHICKENS

*Bring your garden to life and enjoy
the bounty of fresh eggs from your own
small flock of happy hens*

JOHANNES PAUL AND WILLIAM WINDHAM

BARRON'S

First edition for North America published in 2013 by Barron's Educational Series, Inc.

© 2012 Interpet Publishing

First published in 2012 by Interpet Publishing
Vincent Lane, Dorking, Surrey, RH4 3YX, England

All inquiries should be addressed to:
Barron's Educational Series, Inc.
250 Wireless Boulevard
Hauppauge, New York 11788
www.barronseduc.com

ISBN: 978-1-4380-0200-2
Library of Congress Catalog No.: 2012948933
Printed in China
9 8 7 6 5 4 3 2 1

The Authors

Johannes Paul was born in Sussex in 1978 and grew up on a small farm with many animals including pigs, goats, horses, and chickens. After studying Mechanical Engineering at Brunel University, he gained a Masters in Product Design from the Royal College of Art in London. Together with three friends he founded Omlet in order to encourage more people to keep chickens. Omlet produces the eglu—a modern chicken house suitable for a small garden that makes keeping chickens easy and fun.

William Windham was born in Norwich in 1978 and lived in the deepest Norfolk countryside for the next 20 years. He studied Engineering at Cambridge University and Industrial Design at the Royal College of Art. Since graduating he has concentrated on the Omlet company supplying people with chicken houses and a little of the good life. Chickens are central to his day's work—just walking into the studio in the morning involves tripping over at least ten chickens!

Credits
Editor: Philip de Ste. Croix
Designer: Phil Clucas MSIAD
Studio photography: Neil Sutherland and Geoff Rogers
Diagram artwork: Martin Reed
Production management: Consortium, Poslingford, Suffolk
Print production: (TBC)
Printed and bound in China

The authors and publishers would like to offer their thanks to the following people and companies who have provided valuable assistance during the production of this book: Robin Clover for the loan of the HandyHen chicken house pictured on page 10; Forshams Cottage Arks for the loan of the ark pictured on page 12.

For advice and assistance with the breeds photography: Tony, Sue, and Kate Beardsmore, Cathy Burton, Judith Burton, Colchester Poultry Club, James Firth, Nicola Firth, Stephen Flory at The Henhouse Garden Co., David Francis, Lester Frenzel and Daniel, Tim Fuller, Lana Gazder, Simon and Tracey Hayter, Charlotte Heales, Sally Hutton, Lloyd Ince, Mayfield Poultry, Priscilla Middleton, Michael Neve and Joss Parsons at Denmans Garden, Fontwell, Rachel and Andrew Pitt, David Scrivener, Jill Tait, Joy and Steve Thorpe, Mulberry Mill Stud, Suffolk.

All the photographs reproduced in this book, with the exception of those listed below, were taken by **Neil Sutherland** (practical section) and **Geoff Rogers** (page 46 and birds in the breeds section) and are the copyright of Interpet Publishing Ltd.

Jane Burton, Warren Photographic: 1, 3, 44, 45 center.

Fishers Woodcraft: 10 bottom right (both).

Forshams Cottage Arks: 11 top, 11 center right; 12 bottom, 13 center and bottom, 18 top, 22 bottom, 45 top.

iStockphoto.com: 59 bottom right (Mary Lee Woodward), 77 bottom right (Ken Hewitt).

Omlet: 16-17, 17 center right.

Shutterstock.com: 17 bottom left (Richard Pratt).

United States Department of Agriculture: 51.

Wikimedia Commons: 49 top right (Gmoose1)

Contents

INTRODUCTION

An unusual and rewarding pet

Friendly, interesting, and inexpensive to keep, chickens can be kept in most gardens with only simple equipment needed to house them. It is not necessary to have a cockerel; indeed, unless you have very understanding neighbors, it is not advised. Your chickens will lay eggs just as happily without one. As chickens are social animals, they should be kept in pairs or more, and they respond well to human contact. With undemanding care and attention, these intriguing pets will provide you with delicious, fresh eggs and a lot of entertainment.

Chickens have been around for a long time

The first chickens lived around 5000 BC in the jungles of South Asia and were called Red Jungle Fowl. Roosting high up in the trees, they had large talons and strong beaks and were very much like a bird of prey. There is evidence that there were also early types of

chickens in other parts of the world. In South America, the Arauca Indians bred a chicken that laid blue eggs, called the Araucana. Its purity was fiercely guarded by the Arauca and it is still bred today.

As trade between different countries and continents increased, chickens spread across the world, merchants often bringing home unusual and pretty chickens from far away countries as presents for their wives and families.

Left: You don't need a vast estate to keep chickens. Virtually every garden can accommodate a pair.

Domestication and breeding

The Ancient Egyptians developed the technique for hatching chickens artificially, and this was the beginning of the mass domestication of chickens for food and eggs. The Ancient Greeks valued the strength and beauty of the cockerel and it became a symbol of bravery. Used for sport in cockfighting, cockerels were also sacrificed to the gods. Most households kept chickens for their eggs. When Queen Victoria was given some chickens, their popularity in Britain took off. Cockfighting was outlawed, and the Victorians set up clubs and held shows where people could display their chickens in competitions— a tradition which continues to this day.

HANDY HINTS

ROYAL CHICKENS
Queen Victoria was given Cochin chickens, a large, heavy breed from China, which have feathery legs that make them look as if they are wearing trousers. When people first saw them, they were fascinated by these enormous birds, the likes of which they had never seen before.

CHICKENS ARE EVERYWHERE!
The Warner Bros. cartoon character Foghorn Leghorn and the animated film Chicken Run (in which some ill-fated chickens plot their escape from the Tweedy poultry farm) are just a couple of examples of the universal popularity of chickens. Our language itself is infused with chicken references—think about having a hen party, being hen-pecked, feeling broody, having something to crow about, and ruling the roost. And don't count your chickens...!

DIFFERENT AGES, DIFFERENT NAMES
You can call a chicken a chicken regardless of its age or sex, but there are special names as well. Before she lays her first egg a female chicken is called a pullet; after that she becomes a hen. A young male chicken is called a cockerel; when fully grown he is known as a cock.

Below: Domesticated chickens have lived their lives alongside humankind for thousands of years.

5

BASIC ANATOMY

The head

A chicken has an incredible-looking head compared to many other birds. The fleshy comb on the top of the head comes in many different shapes—from big and floppy "single" combs to small and spiky "rose" combs. The comb and the smaller fleshy protuberances under the beak, called the wattles, generally become slightly larger and redder as the chicken comes into lay. With its keen close-range eyesight, a chicken can use its beak to pick up grains of corn with amazing accuracy. But the chicken doesn't chew them; it has no teeth—hence the saying "as rare as hen's teeth." If you look closely at the beak, you will also notice two nostrils near the back edge.

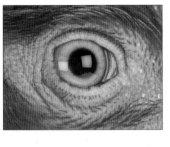

Left: Chickens have an extremely well-developed sense of sight; their vision is more acute than a human's.

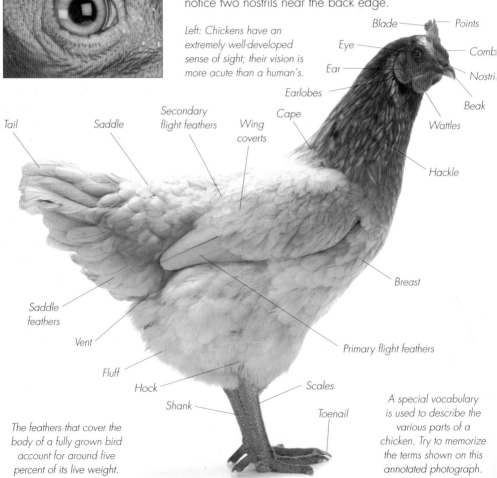

Blade — Points

Eye — Comb

Ear — Nostril

Earlobes

Beak

Secondary flight feathers

Cape

Wing coverts

Tail

Saddle

Wattles

Hackle

Saddle feathers

Vent

Breast

Fluff

Primary flight feathers

Hock

Scales

Shank

Toenail

The feathers that cover the body of a fully grown bird account for around five percent of its live weight.

A special vocabulary is used to describe the various parts of a chicken. Try to memorize the terms shown on this annotated photograph.

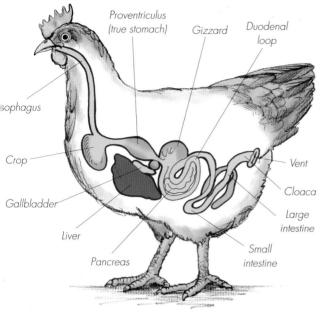

Proventriculus (true stomach)
Gizzard
Duodenal loop
Esophagus
Crop
Gallbladder
Liver
Pancreas
Vent
Cloaca
Large intestine
Small intestine

Digestion

Food is sent on its way down the esophagus by a pointy tongue. The first stop is the crop, a pouch that holds and softens food. If you pick up your chicken at the end of the day, you will be able to feel a roughly golf ball-sized lump on the front of its chest. This is the crop full of the day's food. Next, the soft food passes through the proventriculus, where digestive acid and enzymes are added, and on to the gizzard, a muscular organ where food is ground up. Any insoluble grit the chicken has picked up helps the food to be ground up in the gizzard. The rest of the process takes place in the intestines, where all the nutrients and water are absorbed before waste matter leaves the chicken through the vent (right).

HANDY HINTS

THE BEAK
The beak is made of keratin, the same substance that makes up human fingernails. Occasionally, the top part (the upper mandible) grows too long and overlaps the bottom part. You can trim the top part back with nail clippers and smooth it with a nail file. However, if you are uncertain about how to do this, take the bird to the vet.

Earlobe

HEARING AIDS
A chicken doesn't have external ears; they are located under fleshy areas at the back of the face. Sometimes the color of the earlobe denotes the color of the egg, so if a chicken has white earlobes, it is likely that it will lay white eggs.

EXIT POINT
Chickens have an all-purpose exit hole called a vent. Both droppings and eggs are produced from the same hole. The digestive tract and reproductive system join just inside the vent in an area called the cloaca.

Basic Anatomy

The limbs

Chickens aren't very good at flying. Smaller breeds can achieve some impressive flapping jumps, but the heavier breeds never come close to taking off. The wings are made up of several sets of feathers, the large ones at the front of the wing being the primary flight feathers. They may not be good in the air, but most chickens can run faster than you would expect! A few breeds have feathers on their legs, but most chickens' legs and feet are just covered in scales that provide good protection against all the digging around that they do. At the end of their legs are three toes with nails that they keep short by scratching in the dirt.

Shank

Toenail

Scales

Above: The scales on the shanks are made of specialized skin cells. In hot weather increased blood flow through the legs helps birds to keep cool.

Right: Feathers help to insulate a bird and to aid it in flight. While chickens are not very able flyers, they can usually travel a short distance by flapping their wings, so you must make sure that any outside runs are secure to keep them safely inside.

Secondary flight feathers (18)

Primary flight feathers (10)

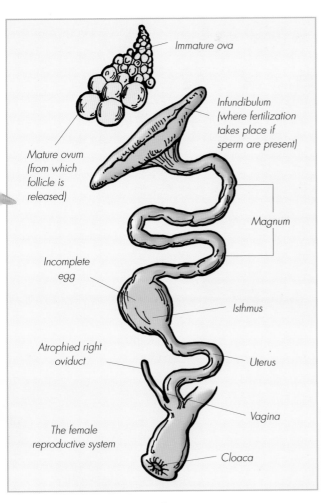

Immature ova

Infundibulum
(where fertilization
takes place if
sperm are present)

Mature ovum
(from which
follicle is
released)

Magnum

Incomplete
egg

Isthmus

Atrophied right
oviduct

Uterus

Vagina

The female
reproductive system

Cloaca

How is an egg made?

A chicken's body contains lots of tiny undeveloped follicles (yolks) which slowly grow one by one to the size we recognize as an egg yolk. When a follicle is ready, a yolk is released from the ovary and travels down a section of the oviduct (the magnum) where it is attached to and coated in egg white. Next, in the isthmus, thin membranes form around the white and the egg then moves down to the uterus where the shell forms. The last stages see color and a protective layer called the cuticle being added. The egg is now ready to be laid! From start to finish the whole process takes around 25 hours.

PREENING
To keep their feathers in good condition, chickens don't only have dustbaths to clean themselves. They also transfer oil from a gland by their tail to their feathers and skin using their beak.

When preening, chickens work an oily substance into their skin.

FINE FEATHERS
Chickens feathers don't just vary in color; there are also striking differences in texture as well. The Silkie has unusual feathers that feel silky and soft and actually look like fluffy fur rather than feathers.

CROWNING GLORY
The comb on the top of the head and the wattles dangling underneath the beak differentiate the chicken from most other birds. They are secondary sexual characteristics and are bigger on a male chicken, a cockerel, than on a hen.

TRADITIONAL CHICKEN HOUSES

The design is important

Chickens' housing requirements are quite simple, as they tend to spend most of their time outside. The house is basically used for laying eggs, sleeping, or as a shelter in bad weather. It is quite possible to build your own house using materials available in most DIY shops, or you can buy a ready-made one. Most houses you can buy are made of wood. They usually come with a run attached to the house. The basic requirements are a perch to sleep on, a nesting box, and security from predators. When building or buying a house, you should also consider how many chickens you are going to keep, how easy it is to clean, and how easy it is to move.

Both right:
Larger chicken houses
are available if you want
to keep more than just two or
three birds. These designs are
suitable for 6–10 birds (bottom)
and up to 15 chickens (top).

Above: Simple wooden chicken houses suitable for the garden should have a secure door which can be locked at night when the birds have gone in to roost. Predators pose a continual threat to chickens.

Security

The house should be secure against all types of vermin including foxes, badgers, and rats. Be sure that there are no easy access holes through which these unwelcome visitors can get in. The door that the chickens use to go in and out should have a strong lock, as a persistent animal like a fox will often work out how to open a simple catch. If there is another door for collecting eggs, make sure that too is secure.

HANDY HINTS

ALLOW VENTILATION

Ventilation holes should be situated near the top of the house. The structure should be draft-free as chickens really don't like wind and rain blowing in on them. Good ventilation also prevents the buildup of ammonia caused by their droppings.

MOVING THE HOUSE

Smaller wooden houses often have handles at both ends so that you can move them from place to place; larger houses may have wheels to make this task easier. It is really useful if the roof can be removed so that you can have maximum access to the inside for cleaning.

Above: With handles and wheels this ark is easy to move around.

CUSTOM CREATION

You can customize a dull wooden house by painting it your favorite color, writing your chickens' names above the door, or maybe training plants to grow against it to make it really blend into your garden.

Traditional Chicken Houses

Perches

The perch should be a 2.5–3.5 cm wide bar that the chickens can grip comfortably with their feet. The edges should be rounded, and it should be positioned toward the back of the house. If there is more than one perch, they should all be at the same height, and not too high, as the chickens can injure themselves jumping off. Some large, heavy chickens may not use a perch but be happier sleeping on the floor. As chickens do about half of their droppings at night, position a removable tray underneath the perches—it makes keeping the house clean a lot easier.

House and nesting box area

Entrance protected by sliding door

Area of enclosed run

Above and left: There are a range of arks on the market that can accommodate different numbers of birds. They feature a traditional enclosed chicken house with a run attached to it. Birds get in and out of the run by means of a door in the front of the house.

Nesting boxes

Your house should have at least one nesting box per four chickens, as they will take turns laying eggs. The nesting box can either be inside or mounted outside the house. You can use straw, dust-free wood shavings, or even shredded paper to line the nesting box. Never use hay as this can develop a mold that can cause your chickens health problems. If there is a door or flap in the side of the house near the nestbox, it will make collecting the eggs much more convenient.

Left & Below: This larger house can hold up to 24 laying hens.
1 *Roof to nesting box area.*
2 *Sliding door.*
3 *Side wall (removable for cleaning).*
4 *Removable tiered perches.*
5 *Bank of nesting boxes.*
6 *Partition between nesting boxes.*

HOUSE PROUD

Remember to keep your wooden chicken house in good condition by treating it once a year with a safe wood preservative, not only to keep it looking good, but to prevent it from becoming weakened by the weather and so liable to rot. Keep the birds far away while you do this and let the stain dry thoroughly.

Above: Cleaning out the droppings from your chicken house is an unpleasant, but very necessary, chore.

KEEPING IT CLEAN

When selecting a chicken house, if possible, choose a design that allows you to take the inside of the house apart and remove all the internal fittings, for regular cleaning. Mites like to live in the cracks at the end of roosting bars and if you can't remove them, they will be difficult to clean properly.

A 21ST-CENTURY CHICKEN HOUSE

The innovative eglu

This is a brand new type of chicken house designed specifically for the first-time owner. It provides everything that two chickens need for sleeping and laying eggs. Inside there are two main areas, the nesting box and the roosting area. When chickens sleep they like to grip onto a perch, like a branch of a tree. The floor is made of wooden bars allowing the chickens to choose where they sleep. They don't need any straw where they roost, but you can put some in the nesting box. The chickens will take turns during the day to lay their eggs, but they like privacy so try not to peek!

Easy to clean

The smooth plastic surfaces of this chicken house are very easy to clean. A soft brush and a household disinfectant can be used to clean the inside—scrubbing the unit once a month will make sure that your chickens have a hygienic home. The roost bars are also removable leaving nasty parasites such as red mites (which love nooks and crannies) nowhere to hide. All the droppings inside fall through the roost bars onto the droppings tray. The tray then simply slides out allowing you to carry the droppings to the compost heap.

Behind closed doors

The eglu keeps the chickens warm in the winter and cool in the summer in the same way that double glazing works in your house. An air gap between two layers keeps the inside temperature constant. The front door can be opened and closed using the handle on the top. It locks too, so at night the house is a cozy, secure retreat for the chickens. In the morning when you open the door, the chickens will come rushing out looking for breakfast. You might find yours too, freshly laid in the nesting box!

HANDY HINTS

MOVING HOUSE
When you go on vacation and need to take your chickens to a friend's house, it is simple to transport the chickens securely inside the unit. Try to keep traveling time to a minimum, especially when it is hot. Chickens don't mind being shut in for a few hours at a time, but try to keep stress to a minimum.

LET THE LIGHT IN
Once you have cleaned the inside, leave the lid off for a few hours to let some sunlight in. The UV rays will kill off many microorganisms that prefer hiding away in dark, moist places. It pays to do this about once a month.

NO MAINTENANCE
Unlike a wooden house, this design doesn't require any long-term maintenance. Wooden houses need to be treated or painted every year, but plastic only needs regular washing and cleaning.

LARGER HOUSES

Think ahead

If you plan to keep more than just a few chickens, you will need to buy a larger house that can accommodate them comfortably. Overcrowding among birds causes them stress and may result in a greater incidence of illness and disease. Commercial manufacturers may say that a house is designed for a certain number of birds, but bear in mind that often they are thinking of commercial layers which are relatively small-bodied. If you plan to keep much larger breeds, you must take this into consideration. Allow around 10 square feet (1 sq m) for every five birds. Also consider whether your birds will be kept on a free range or confined to a run. If they are in a run, they will spend more time in the house as they will not find trees and hedges under which to shelter in bad weather.

Modern designs

Building on the popularity of the original eglu, the eglu cube is a flexible chicken house in which between five and ten chickens may be kept. It has twin-walled insulation and draft-free ventilation to keep the chickens warm in winter and cool in summer. Available as a stand-alone house, it will sit comfortably in an existing run or fenced-off area such as an

orchard. It is also available with an integrated run, which provides an easy-to-move, secure area. The standard run is 6.5 feet (2 m) long, but can be extended with 3 foot (1 m) sections depending on the number of chickens that will be kept. Inside, the birds have a large roosting area fitted with a slatted floor. All the droppings fall through the slats onto slide-out dropping trays, which can then be emptied onto the compost heap.

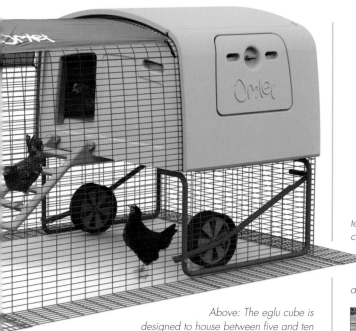

Above: The eglu cube is designed to house between five and ten chickens. The wire run offers protection from predators.

HANDY HINTS

PLENTY OF SPACE

You need a big garden if you are going to keep as many as ten chickens. If you have not kept chickens before, it's a good idea to start with just two or three to see how you get along with them. You can always add more at a later date.

Above: A community of chickens in your garden can quickly become a magnet for children.

BIGGER BREEDS

If you are interested in keeping bigger breeds, such as the Jersey Giant, Buff Orpington, or Brahma, you're going to need a bigger house. Don't stock the house to its maximum advertised capacity. Your birds will appreciate a bit of extra legroom.

Above left: This traditional wooden house offers accommodation for up to 12 laying hens or six larger fowl. It features three separate nesting boxes located on the side of the house. The side walls and perches can be removed for easy cleaning.

Above: The more chickens you keep, the more you will have to plan ahead with regard to secure housing and a safe area to range.

WORK LOAD

More chickens of course mean more cleaning out. You will have to clean out a larger house more regularly if you keep as many as ten chickens in it.

THE RUN

Small runs

Depending on whether you have a small urban garden or a large garden in the countryside, you will probably have a different way of keeping your chickens. Many chicken houses come with a run attached in which some people keep their chickens all the time. Others use it as a secure area to keep the chickens in when they are not around. If you are going to be keeping your birds in a run for most of the time, your chickens should preferably have more than 10 square feet (1 sq m) of space each.

Above: Chickens need time to scratch around in the outside world. A run affords them security as they do so

Mesh skirt to deter tunneling predators

Above: The eglu is provided with a secure wire run that attaches to the front of the chicken house.

Large runs

An alternative to keeping a small run and letting your chickens out to roam anywhere they please is to section off an area of the garden around their house. If you decide to do this, make sure that it is properly secure. Remember that foxes can climb so don't forget a roof! The grass will soon wear thin if you keep the chickens in one place so putting bark chippings down is a great alternative. A small run can be moved to a new area of grass to keep it fresh and disease-free while bark chippings can be simply raked out and a new layer spread every month or so.

Protecting your run

There are several ways to stop animals from tunneling into your chicken run. You can excavate and bury the netting into the ground so that any tunneling attempt is thwarted. However, this has the drawback of preventing the run from being moved around easily. Alternatively, you can construct your run with netting on all sides including the bottom, although this makes the floor area more difficult to clean. Or you can make your run with a "skirt" of meshing not less than

Above: Feeders and drinkers can easily be situated in a run so that the chickens have ready access to food and water whenever they need it.

7 in (20 cm) wide all around. This stops predators like foxes from digging under the base of the run wall while still enabling you to move the run easily and keep the birds protected.

Below: A house with a run attached to it can be moved around the garden—your grass will appreciate this!

HANDY HINTS

ADD INTEREST
You can do many things to make life more interesting for chickens that stay in a run most of the time. Hang up old broccoli stalks for them to peck at. Don't hang them so that they have to jump to reach them, though, as this could upset their egg laying. If you put a mirror up, your birds will spend a surprising amount of time looking at themselves.

Broccoli makes a welcome treat.

MESH SIZE AND MATERIAL
Make sure that the size of the holes in the wire run is sufficiently small, preferably no bigger than 1 in (2.5 cm), so that your chickens can't stick their heads out. For a stronger run, try using "weld mesh" instead of conventional chicken wire. It is made from thicker wires welded together. When choosing a chicken house and run, be aware that if the sides of the run are not vertical, the chickens cannot really use the area at the edge of the run without rubbing their combs against the wire.

FOOD AND WATER

Chickens like variety

Instinctively good at foraging, chickens are quite capable of finding a good proportion of their daily intake of food if they are given enough space and a variety of surroundings. They will eat grass and other greenery, worms and insects, as well as find water in empty flowerpots and puddles. However, laying an egg takes a lot of energy and so it is a good idea to provide a laying hen with the correct balance of protein, carbohydrates, vitamins, and minerals, especially if she is not allowed to roam free range. The best way to ensure she gets proper nutrition is by using a specially prepared feed for laying hens.

Be consistent with the feed

A fully grown chicken needs about 4.5 oz of feed a day. Chicken food for laying chickens is called layers mash or layers pellets. They are a mixture of wheat, barley, oats, and maize. A good quality feed will contain soya and not other forms of protein, and it should say on the label that it is vegetarian.

Layers pellets are the same as mash except that they have been formed into neat cylinders and are very easy for the chickens to eat. If your hens are confined to a run, it is better to feed them mash as they will find foraging for this more interesting and will normally split up their feeding times with other activities. You should store the food somewhere sheltered so that it doesn't get wet.

Grit

Pellets

Mash

Clean water is important

It is very important that your chickens have a supply of fresh, clean water. You should refill the container every day, and in especially hot or cold

Above: These fountains work on a vacuum principle that maintains a constant water level.

weather, check it at least twice a day to make sure that the hens have water to drink. On a hot day, a single chicken can drink around 16 oz (475 ml) of water. You can use either plastic or metal containers—look for ones that prevent the chickens from being able to tip the water out or stand in it, and make sure that they are durable and easy to clean.

Below: Chickens cannot swallow liquid in the way that we do—they have to tilt back their heads to drink.

HANDY HINTS

EXTRA TIDBITS

You can supplement your hens' diet with scraps from the kitchen. Leftover cooked pasta and rice, as well as vegetables and fruit, are usually enjoyed. The best option is to experiment with different scraps to see what they like, as they can be quite fussy. Avoid feeding them any salty, sugary, or fatty foods, citrus fruit, or meat.

Pasta

Broccoli

Rice

GRIT FOR GRINDING

Chickens don't have teeth and can only digest food by first grinding it up in their gizzards using small stones that they pick up from the ground. If they can't find stones in their surroundings, you must supply them. Leave a container of grit in the run for the chickens to take what they need.

FOOD FOR YOUNGER CHICKENS

If your chickens are not point of lay (old enough to start laying) then you will need to feed them a different feed called growers pellets or mash.

PREPARING YOUR GARDEN

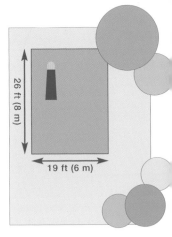

Size matters

When you start thinking about getting chickens, you must make sure that the number of birds you buy is right for your garden. However much you want to keep chickens, a window box is not suitable! As a rough rule, around 540 square feet (50 sq m) for two birds is sensible, so it is quite possible to keep chickens in many city gardens. With this much space and a mixture of ground types, your chickens and garden plants will happily coexist.

Above: An area of around 516–538 square feet (48–50 sq m) in a garden is ample space in which to keep a couple of chickens.

Suitability

If your garden has perfectly manicured lawns and tidy flowerbeds, you must prepare yourself and the garden for a little bit of messiness. So what do chickens like in the garden? They love areas with bushes and shrubs to hide and shelter under. Chickens clean themselves by making a dustbath, so your chickens will enjoy areas of loose earth, such as the back of a flowerbed. If there is nowhere obvious to take a dustbath, they may well scratch a bath

Above: Hens are opportunistic feeders enjoying worms, insects, seeds, and greenery as they forage.

into your lawn—be warned! Chickens are not housetrained so they will poo all over the place. Chicken manure is great mixed into compost, but can be too harsh and caustic if put straight onto plants.

Left: Chickens and pristine lawns do not go together. An area of rough grass, like this, is a better option.

Above: Chickens offer their owners much more than just free eggs; they also bring a garden to life as they scratch and putter around in the beds.

Chicken proofing

A 6.5 foot (2 m) fence will keep them in nicely, but remember that although chickens don't fly, they can perform some impressive jumps, so anything left near the fence will enable them to hop up and over. Chickens love to nibble at new shoots and can be a real pest in the vegetable patch. To prevent this, you can section your precious seedlings off with some flexible fencing or use smaller wire cages to protect individual areas. Encourage the chickens to make a mess where you want them to by giving them an area of rough ground with bare earth to forage in.

Above: Keen gardeners may decide to erect some protective netting to keep their birds away from any prized plants.

HANDY HINTS

HAZARDOUS CHEMICALS

If you must use weedkiller or other chemicals, such as slug pellets, in the garden, make sure that your chickens are kept far out of the way. Don't let them back into the treated area until you are absolutely sure that it is free from these potentially lethal hazards.

FLOWERPOT DUSTBATHS

If you have a small garden and don't have anywhere for your chickens to dustbathe, a large flowerpot full of sandy earth can do the trick. Not only does it allow them to clean themselves, but it is also very amusing to watch.

WATCH THE WATER

Chickens can't swim! Their feathers tend to soak up water rather than causing it to run off like it does from a duck's back. Chickens will probably just drink from a pond rather than trying to swim, but it pays to keep your eye on them the first few times they approach one.

BUYING A CHICKEN

Where to find chickens

It is unlikely that your local pet shop will sell chickens. Try instead the classified ads in your local paper, specialist magazines, or ask your vet or a local supplier of chicken feed. If possible, it is always better to go and see the chickens for yourself in their surroundings and ask the seller any questions you might have. Choosing chickens that have been reared together and are of the same age should ensure that they are already good friends and unlikely to fight. Before you proceed forward, you should decide how many chickens you want, whether you can keep a cockerel, and if you want a chicken that will be easy to look after and lay lots of eggs (a hybrid or crossbreed) or a pure breed that looks exotic but will require more attention.

What to look out for

A healthy chicken will have bright eyes, glossy feathers, and be perky and active. A pullet will not yet have a big red comb (this will develop when she begins to lay). The legs should be smooth and the breast firm. You should check the bird's beak and avoid any that have a discharge; watery eyes and a dirty vent area are also signs that the bird is not 100 percent fit.

Glossy feathers

Active and alert demeanor

Bright eyes

Beak clean, nostrils free of discharge

Firm breast

Is vent area clean?

Legs should be firm and smooth

Above: It is a good idea to visit the breeder and check a bird over before you decide to buy it.

Right: One way of establishing whether a chicken is in lay (i.e., mature enough to lay eggs) is to measure the distance between its pelvic bones, as illustrated here. A bird that is in full lay should exhibit a span that is three fingers wide between its pelvic bones.

Below right: A pullet is a young female chicken that has not yet laid an egg. Pullets typically have smaller combs than mature females.

HANDY HINTS

LOOKS OR EGGS?
Don't just go for the prettiest bird you see—often the chickens with the most amazing plumage don't lay very well. Remember as well that chickens with feathery legs will get muddy and you will have to spend time cleaning them.

WHO ARE THE PARENTS?
If you are buying a pure breed hen, ask to see the parents. It will give you a good idea of what your chickens will look like when they are adults and whether they will have a friendly, docile character.

What age to buy

Ads may describe a chicken's age as being at point of lay—often abbreviated to POL. This means that the chicken is about to start laying but doesn't necessarily mean that this will be the chicken's first laying period. You should always ask the age of the chicken as some people will sell their chickens after two years to make room for new hens. A chicken that has not yet laid an egg is called a pullet. It is possible to tell the sex of some crossbreeds at just a day old, while others can only be reliably sexed when they are at least four weeks old. Ideally, buy your chickens when they are at least 16 weeks old, as at this age they can still be tamed and there can be no doubt as to whether or not they are male or female.

THE FIRST DAY AT HOME

Bringing your chickens home

You can transport your hens home in a spacious cardboard box, lined with straw and with holes cut in the side for ventilation. Alternatively, you could use a wire pet carrier such as you would use for a small dog. Make sure that the container you use has plenty of ventilation; you can cut holes through a cardboard box for this purpose but make sure you do it before you put the chickens in the box! On long journeys check that the chickens don't get too hot; warning signs are panting and trying to cool down by spreading their wings. You should stop every couple of hours to give them some water.

Other pets

As a general rule chickens get along fine with other pets, such as cats and dogs. Try to let your chickens settle in for a few days before introducing them to other family pets and make sure that you only introduce them gradually when they are safely inside their run. Initially the chickens may flap around making them even more interesting for an inquisitive dog, but given time they will usually get along fine in the garden, puttering around with a healthy respect for one another.

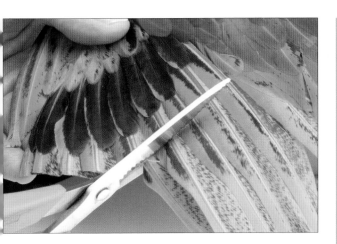

Wing clipping

Although most chickens are unable to fly high or over long distances, it is quite common and painless to clip their wings. It is only necessary to clip one wing to unbalance the chicken and prevent it from being able to fly onto or over fences. The person you bought your chickens from may have already done this for you. Alternatively, ask someone to help you by holding the chicken while you trim the long primary flight feathers on the edge of the wing. Using scissors, cut the first six or seven feathers back to the point where the next line of feathers starts.

Above and top: Wing clipping is a simple and painless way of preventing your chickens from fluttering out of an open run or coop. Just clip the primary feathers on one wing, so that the bird is unbalanced in flight.

HANDY HINTS

FROM BOX TO RUN

If you are unsure about handling your chickens, open the box you brought them home in inside their new run. This guarantees that you won't spend the afternoon chasing chickens around the garden if one leaps out of the box unexpectedly.

BEDTIME

The first night your chickens may not know where they are supposed to sleep. To solve this, try putting a torch inside their house. They should be drawn to the light and once they are inside you can close the door and take the torch out.

THE FIRST FEW DAYS

To make sure that the chickens know where to roost, keep them inside the house and run for around five days. That way, if you let them out into the garden, they will have learned that the coop is a nice, safe place to roost and will come back to roost at dusk.

EGGS

Eggs are full of goodness

The eggs from your own chickens are really fresh and a great source of energy. An average egg contains about 70 percent water, 10 percent protein, 10 percent fat, and 10 percent minerals. They contain all eight of the essential amino acids, plus vitamins A, B, D, and E to keep your body healthy. The amount of protein in a single egg is equivalent to 14 percent of the recommended daily allowance for an adult. The whole egg contains about 75 calories, with around two-thirds of this contained in the yolk.

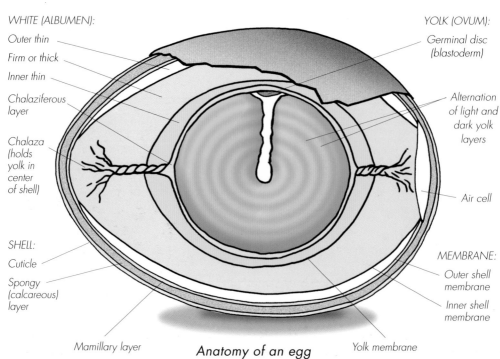

WHITE (ALBUMEN):
Outer thin
Firm or thick
Inner thin
Chalaziferous layer
Chalaza (holds yolk in center of shell)

SHELL:
Cuticle
Spongy (calcareous) layer
Mamillary layer

YOLK (OVUM):
Germinal disc (blastoderm)
Alternation of light and dark yolk layers
Air cell

MEMBRANE:
Outer shell membrane
Inner shell membrane
Yolk membrane

Anatomy of an egg

Size and shape

The first eggs that your chickens lay might well be surprisingly small—even as small as a grape. They will get bigger as the chicken's body adjusts and develops until she is consistently laying eggs weighing about 2 oz (56 g). Your chicken may well lay some rather peculiarly shaped eggs over the years, from long, thin eggs, to ones that look more like crumpled-up balls of paper! The cause of the irregularities could be the chicken being frightened or just old age.

Eggs—one of nature's most amazing feats of packaging.

Number of eggs

The number of eggs that your chickens will lay depends on the breed. Some crossbreeds can lay over 300 eggs a year whereas the most a pure breed hen will lay is around 250. Because chickens are sensitive to the number of hours of daylight, they lay more eggs in the summer than in the winter. If you have two crossbreed hens, you will get around 12 eggs per week in the summer whereas in the winter this will reduce to eight eggs. A drop in the number of eggs your chickens lay is also one of the first signs that they aren't completely happy.

HANDY HINTS

COOL STORAGE
Eggs don't need to be kept in the fridge. If you keep them pointed end down in a cool place like a larder, they will last for up to three weeks. If you are not sure whether an egg is fresh, put it in a bowl of water; if it sinks it is fresh, if it floats it is rotten.

A rotten egg

A fresh egg

YOLK COLOR
The color of the yolks in your chickens' eggs depends on their diet. The more greens they eat, the yellower the yolk will be. Some other foods affect the yolk's color; for instance, a few acorns can give it a greenish tinge.

SOFT EGGS
Weak shells can be caused by a lack of calcium in your chickens' feed. You can buy special poultry grit to remedy this, but an alternative is to bake old eggshells, grind them up, and mix these into the daily feed ration.

CHICKENS' DAILY ROUTINE

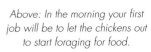

Morning

Your chickens will have had a completely inactive night simply holding on to a perch with their eyes shut. As soon as it gets light, they will be awake and ready to be let out to start feeding. Assuming your chickens are more than about 18 weeks old, they will probably lay

Above: In the morning your first job will be to let the chickens out to start foraging for food.

an egg in the next few hours. If they live in a house with a nestbox, they will make themselves comfy in it, then stand up 20 minutes or so later having laid an egg. If you let your chickens out free range, then the next part of your day will consist of a hunt to find the eggs!

Daytime

Chickens will eat most of their food in the morning, so make sure that they have plenty of layers mash to fill up on before you give them any treats. Eating will be followed by a good preening session. A dustbath will allow sand and earth to filter through their feathers, dislodging any mites. If it's a hot, sunny day, they may well lie outside, looking like they've melted with one wing outstretched! The rest of the day will be spent puttering around, scratching for bugs and tasty morsels.

Left: Chickens that are contained in a run will normally disappear back into the nestbox area to lay an egg.

Above: If you keep two chickens, you can normally expect to harvest two eggs a day from them during the summer season.

End of the day

A very useful habit of chickens is that they never fail to go and roost at dusk. Make sure that you remember to leave the chicken house door open in the evening. If they are ready to go to roost and find they can't get in, they will make a bit of noise and then find alternative accommodation. This is likely to be an unprotected branch which will leave them vulnerable to ending up as a fox's nighttime snack.

Left: If you allow your chickens to wander around free range in the garden, you must be prepared to search around to find their eggs, as they will not necessarily use the nesting boxes in the chicken house for laying.

SEASONAL TIPS

Keep them cool, keep them dry

Most breeds of chickens are hardy and can live happily outdoors all year round. However, you must provide them with shade and shelter from the sun, wind, and rain. The different seasons will affect how active your chickens are and the number of eggs that they lay. Although they will be less active in the winter, they will make the garden a brighter place all through the year as they provide visual interest and attract other wildlife.

Above: Chickens perform a useful job in the garden picking up pests, but they can be tough on young plants.

Spring and summer

Chickens are sensitive to light and so in the spring and summer they will rise earlier and go to bed later. The longer daylight hours will also encourage them to begin laying if they have taken a break from this over the winter. On long, hot days chickens will be thirsty, so check regularly that they have enough water; you could put extra containers out, too. If you let your chickens roam in the garden, they will pick off pests like slugs and flies. You may want to protect young shoots, however, as chickens can be quite partial to tender young plants. Depending on the breed of chicken, they may go broody in the spring or summer; to prevent this make sure you collect the eggs regularly.

Autumn and winter

Chickens, protected by their feather overcoats, don't mind the cold, but they prefer not to get wet. As the weather turns, your chickens may spend more time under cover. If they can't find this outside, they will stay in their house. Covering the run to keep them dry and shaded from the wind will encourage them out to feed. You should check their water regularly for freezing, and use an old string bag to hang greenery in the run for the chickens to peck at. You will get fewer eggs in the winter because of the shorter days, and some pure breeds may stop laying altogether. In very cold weather it is a good idea to rub vaseline into the chickens' combs to protect them against frostbite.

HANDY HINTS

LONGER NIGHTS

Don't forget that as the days get shorter, your chickens will go to bed earlier and you will need to shut the door to their house earlier.

MIND THE MUD

Some pure breeds with long feathers will get very muddy, and their feathers can get damaged if they are kept in muddy conditions in the winter. To prevent this, keep them on a surface that drains well, like bark chippings.

SAFER INDOORS

If you or your neighbors like to celebrate special events with a firework display, it would be best to bring your chickens into your house. You can put them in a spacious cardboard box lined with straw and with plenty of air holes and then pop them back into the garden in the morning.

GOOD MANURE

Chicken droppings can be put on the compost heap where they will help to make a very rich compost mixture that will improve the soil in your garden. The manure can even be put on some plants directly—it is especially good for currant bushes—but take care; it is too strong for most flowers and will burn them.

Left: In freezing conditions, a dab of vaseline rubbed onto the comb will protect it from the cold.

33

KEEPING THINGS CLEAN

Weekly care

Weekly care is principally based around cleaning the chicken house and making sure that there is not a buildup of droppings. Remove and replace any soiled straw or shavings and scrape out any droppings from the house. This is much simpler if your house has a droppings tray of some variety. Putting down some sheets of newspaper or cut-up cardboard boxes stops the droppings from sticking to the floor, making it easier to lift the soiled mess out. Also make sure that your feeders and drinkers are thoroughly cleaned out to prevent any bacterial buildup.

Chopped straw *Wood shavings*

Both left: If your house has a removable droppings tray, it is a simple job to slide it out and hose it clean. Remember that the droppings make a valuable addition to the compost heap where they can rot down.

Disinfecting and mite powdering

However clean you routinely keep your chicken house, it is important that you perform a really thorough clean about every four months. First, do your usual weekly clean. Now take out any loose parts, roosting bars, droppings trays, etc. Using a pet hutch disinfectant available from pet shops, make sure that no crevice is left unscrubbed. Mites like to live in moist, dark crevices. Now leave everything to dry in the sunlight; this kills many organisms not normally exposed to UV. Finally, give the house a good dusting with mite powder. If you have a particular problem with mites, you should liberally dust your chickens around the neck and vent as well. This should ensure you a problem-free chicken house for several months.

Wooden houses need dusting periodically with mite powder.

Above: Perches and roosting bars need a brisk scrub from time to time.

House maintenance

Wooden houses need to be treated with preservative to keep them from rotting. Make sure your chickens are kept away while you are painting. First, check that none of the house has rotted and needs replacing, then follow the instructions on the packaging. Try to treat your house in the morning so that as long a period as possible can elapse before the chickens need to go back inside. If you have a plastic house, maintenance is much simpler and basically just involves a thorough cleaning with soap and water.

Above: Chopped straw is a convenient material with which to line the nesting box. It is easy to spread, and easy to dispose of when soiled.

HANDY HINTS

FOOTBATHS
Quite often your chickens can accumulate mud on their feet that can turn into hard balls. Soaking them in warm water will cause the mud to fall away, making your chickens much more comfortable.

NEST MATERIAL
You can use straw or wood shavings as nesting material. If you use wood shavings, however, make sure that they are manufactured specifically for the purpose. The wrong sort of shavings may contain very fine dust that can cause respiratory problems among poultry.

FEATHERED LEGS
Light-colored birds or long-feathered breeds such as the Pekin bantam can get extremely dirty in the wet winter months. To clean the feathers, use warm water and a little soap. Make sure you rinse them well and gently dry with a towel and some mild heat—not too hot though, as you can damage the feathers.

35

HOW TO HANDLE A CHICKEN

A handful of treats gains their attention.

Building up trust

The first few times you let your chickens out of their run, do it in the early evening so they won't stray far. Stay in the garden with them and entice them to come close to you by scattering treats like sweet corn and raisins on the ground for them to find. You can encourage them to come to you by talking to them and holding out food in your hand. When your chickens are ready to eat from your hand, then you can try to stroke them. Avoid fast or sudden movements and gently stroke the feathers on their chests and back.

Above: Feeding by hand allows chickens to grow confident of being around you.

Picking a chicken up

The next step is to pick your chicken up—be confident and bring both hands down over her back holding the wings against her body so that she can't flap. Lift her from the ground and hold her against your body, supporting her underneath with one hand; she may struggle a bit but should, once she feels safe, be quite happy to let you hold her.

Left: Try to pick up a bird for the first time while she is calm and unflustered by your presence.

Use both hands to take a firm hold of the body and to stop the wings from flapping.

Difficult to catch

Try not to fluster your chickens by making wild sweeps to catch them. Chickens can move quite quickly and if you try and grab your chicken in a desperate lunge you may end up pulling feathers out. If it is proving difficult, leave them alone for a few minutes to calm down before trying again later. Sometimes it is helpful to have more than one person present so that, between you, you can block off all the escape routes more effectively!

Right and below: Once you have picked a chicken up, hold it against your body using one hand to support it from underneath. In this position, it is relatively easy to keep the chicken quiet.

This "head reversed" position is also very effective.

HOME TO ROOST

Initially you may find it simplest to handle your chicken when she has gone to bed. Open the door of the house and lift her off the perch. Put her back gently. You both should find this less stressful.

Above: Tame birds can pop up in the most surprising way!

AN OPEN INVITATION

When your chickens have become tame you will find that they follow you about in the garden and may even hop onto your lap or shoulder if you sit down. If you leave the back door open, you may even find that your chickens pop in to have a look around your house.

CHICKEN BEHAVIOR

Below: Every henhouse will have a pecking order that establishes the hierarchy within the group of birds. New introductions or young birds have to find their place.

Pecking order

Any group of chickens will have a pecking order. If there is a single cockerel amongst the group, he will naturally be in the top spot with the females forming an orderly queue behind him. If there is no cockerel, then the chickens will work out their own order with what can be vicious pecking and squabbling. The bigger the group, the more complicated and drawn out this process will be. Unless one hen is being badly picked on, do not interfere as—even if it takes a few days—they will finally sort it out and go on to live in harmony with each other. Introducing a new bird or taking a chicken away can trigger the whole process again.

Broody hen

From time to time your hens may go broody even if you don't have a cockerel. You can tell if a chicken has gone broody because she will stay sitting in the nesting box and be quite grumpy with you if you try to move her. It takes 21 days to incubate eggs but even if you take any eggs away from a broody chicken, it is not easy to snap her out of this spell. You should make sure that she has access to food and water, but if she doesn't come out once a day to feed and drink, you should lift her off the nest and put her outside.

Above: When a chicken is about a year old, she will start to lose her feathers. Don't panic—she is not ill; molting is a natural process of feather replacement that takes about six weeks.

Molting

Once a year your chicken will molt. She will lose a lot of feathers from all over her body, starting from behind the head, moving across the back, and then down across the wings. She will look quite bald and then like a porcupine as the new feathers begin to grow through— they look like short, hollow tubes that you can see poking out through the skin. She will not lay any eggs during the molt and it will last, on average, about six weeks.

HANDY HINTS

MAKING A HEALTH CHECK

Keep your eyes open

Spending time with your pet is the first step toward ensuring that it stays healthy and happy. Just a few minutes every day spent observing your chickens is the best way for you to learn about their behavior, habits, and routines. Knowing how your chickens behave normally will make it easier for you to tell when one is ill or in discomfort. Pick your chickens up regularly and give them a quick health check. If you do this daily, you should spot any problems at an early stage.

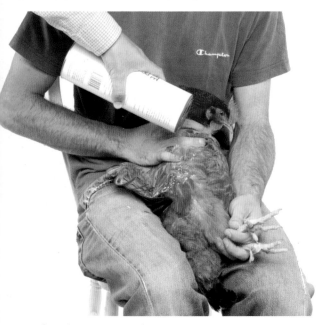

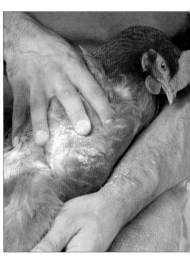

Above and left: Lice are biting parasitic insects that live on the chickens themselves. They can be treated by dusting with a proprietary powder.

Red mites

Red mites like to live in the small gaps and crevices that they find in wooden houses and are one of the most common problems encountered with chickens. They do not live on the chickens, but come out at night and crawl onto them to feed on their blood. The effect on the chicken will be that she looks anemic and stops laying. During the day you should check under the perches for mites. If present, they will be red, slow-moving, and about 1 mm long. You must treat the house with mite powder, dusting it liberally over all the surfaces, paying particular attention to crevices and corners.

Right: Part the feathers to keep an eye open for the presence of lice or mite bites.

Look out for lice

Lice live on the chickens, in litter on the floor of the house, and sometimes in a chicken's dustbath. You can check for them by parting the feathers around the chicken's vent and under her wings where you may see light brown-colored lice and clumps of white eggs. They irritate the chicken and can cause feather loss. The treatment is to dust the chicken, the house, and the dustbath liberally with lice powder until the problem clears up.

Scaly leg mites

If you look closely at a chicken's feet and legs, you will see that they are covered in scales. These should be smooth and clean. There is a mite that lives under the scales that causes them to lift and come off. If you notice this, you can treat it by dipping your chickens' legs in rubbing alcohol twice a week until the problem is cured.

Left: Check your birds' legs regularly, as scaly leg mites can lead to lameness.

LOOKING GOOD

The chicken's body should be firm but not fat, and her feathers should be shiny and complete; any bare patches may be a sign of lice or a skin problem. Most health problems with chickens arise from poor hygiene in the house or from keeping too many chickens in too small a space.

WEIGHT WATCHING

Most chickens are good at regulating their weight and will not overeat. However, if you feed your chickens too many treats, they will not get enough exercise foraging for food, and will become lazy and not lay as many eggs!

FREE RANGE

If you clean the house regularly and allow your chickens access to fresh areas of ground in which to forage, you should rarely have any health problems to worry about.

HEALTHCARE

Worms

Chickens, like many animals, can pick up the eggs of worms when feeding that then hatch and live inside their intestines. Chickens that need worming will maintain or increase their food intake while at the same time laying less or even stopping egg production completely. Their combs will look faded—pink rather than red in color—and their droppings may be runny. If you suspect that your chicken has worms, you should buy a worming treatment from your vet to mix with her feed. Some people worm their chickens twice a year just to be on the safe side.

Left: The crop is an enlarged area in the esophagus where food is stored before it passes into the proventriculus. If the crop feels hard, it is possible that it contains a blockage.

Crop bound

Eating long grass is not good for your chickens and may result in a blockage in their crops. The crop will be visible as a large bulge on the chicken's chest. This will normally only be noticeable in the evening after a full day's eating. If you notice the bulge in the morning and your chicken is not eating, pick her up and feel the crop. If it is hard, then you may pour a little vegetable oil down her throat and massage the lump to loosen and break it up. If this doesn't work, then it is best to take the bird to the vet.

Chickens can catch colds

Check that each bird's beak is clean and the nostrils are clear. Chickens can catch infections from wild birds and any discharge from the nostrils could be a sign that your chicken has picked up a cold. You may also hear her sneezing or see her opening her beak to breathe, a sign that the cold has affected her respiratory system. If you notice any of these symptoms, you will need to take her to a vet for an examination; a short course of antibiotics usually clears up any respiratory problem.

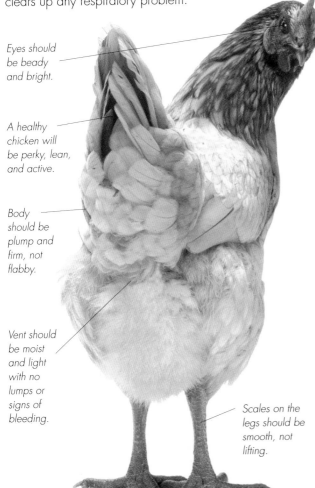

Comb will be bright red when chicken is in lay.

Eyes should be beady and bright.

A healthy chicken will be perky, lean, and active.

Body should be plump and firm, not flabby.

Vent should be moist and light with no lumps or signs of bleeding.

Scales on the legs should be smooth, not lifting.

TELL-TALE SIGNS
Chickens' droppings should be firm with a white cap. Because chickens eat a large variety of things, the droppings may occasionally become runny. However, if you notice that your chicken has dirty feathers around her vent there may be something more serious wrong with her. In such circumstances, a trip to the vet is a wise precaution.

SMELLY BREATH
Putting a clove of garlic in the drinking water can help to keep your chickens fit and healthy. If the chickens stop drinking from the water container, you can try putting a teaspoon of dried garlic with their food instead.

DON'T BUY CHEAP FEED
Using a good quality feed and always having a fresh, clean supply of water are essential to maintaining a healthy pet.

BREEDING CHICKENS

Hatching chicks with a broody hen

In a chicken run with a cockerel, the female chicken will lay fertilized eggs. In the spring or summer a hen may go broody and lay a clutch of eggs, between nine and 15 is normal, before sitting on them for the next 21 days. You should keep a broody hen in a separate house from the rest of your chickens. During this time she will turn the eggs regularly and move them about to ensure that the eggs are all getting an equal amount of heat. She will only come off the nest for about half an hour every day to feed, drink, and take a dustbath.

Below: If you decide that you want to breed from your birds, you will need a cockerel to fertilize the hen's eggs. When she has laid her clutch of eggs, she will incubate them for around three weeks before the chicks hatch out.

Left: A broody hen should be kept apart from the remainder of your flock of chickens. A small garden ark like this is a convenient method of isolating the bird.

Hatching eggs using an incubator

Even if you don't have a cockerel, you can buy fertilized eggs at a low cost. They can be stored for up to two weeks and then be hatched under a broody hen or by using an incubator. An incubator is an artificial way of hatching eggs; it works by mimicking the action of a broody hen by maintaining a constant temperature and turning the eggs automatically every couple of hours. You should still check the eggs in an incubator regularly. It usually takes slightly longer than 21 days for the eggs to hatch.

When the chicks are born

If the chicks have been hatched by a hen, she will instinctively know how to look after her young. However, don't return the mother and chicks to the rest of your flock until the young birds are eight weeks old. If you have used an incubator, you will need to move the chicks to a brooder; this is a small enclosure with an infra-red lamp for warmth and food and water containers. After five or six weeks feathers should start to appear and you can take the chicks out of the brooder, but you must still keep them inside for another two weeks.

COCKERELS WILL NEED HOMES

If you have not kept chickens before, remember that you will need the right equipment to hatch the eggs and to look after the chicks when they have hatched. It is difficult to hatch chicks successfully, and some of the newborn chicks will be cockerels, for which you will need to find homes.

CHOOSE AN INCUBATOR CAREFULLY

It takes a lot of care and attention to hatch chicks successfully, especially if you are using an incubator. There are lots of incubators available and you should always read the instructions very carefully before selecting the right one for your circumstances.

FEEDING CHICKS

Chicks can live for the first 24 hours without any food or water, but after this time you must give them special food called chick crumbs. Water should be provided in a suitable container that they can't drown in. If you have other pets make sure that they can't reach the chicks. If they are outside, keep them under netting so that they are not at risk from predatory birds like magpies.

CHICKEN BREEDS

Introduction to four main breed types

This section of the book introduces readers to a selection of the many breeds of poultry that are kept by chicken fanciers around the world. Some will be familiar names, while others are perhaps less well-known. They have been selected to show some of the extraordinary variety of chickens that are available to pet keepers. Some are especially prized for their egg-laying potential and for their affectionate natures. Others, particularly some of the True Bantams, are supremely elegant birds that are kept primarily for ornamental purposes and for exhibiting at poultry shows. If you get bitten by the showing bug, you may find a whole new world opens up to you as a supremely fulfilling hobby. But for those less-competitive souls who simply want to keep a few happy and productive chickens in their garden, there are plenty of breeds to attract you here, too.

Above: Chickens come in the most beguiling range of colors and plumage types. Not only do they provide a plentiful supply of fresh eggs, they are an ornament to any garden as well.

LIGHT BREEDS *page 48*

These are generally the laying breeds that have provided eggs for the table for centuries. These birds are bred to be productive egg-layers. They do not convert energy from food into fat and so are smaller-framed than fowl destined for the table.

HEAVY BREEDS *page 62*

Derived from large Asian poultry stock that were bred to be big and plump for the table. Heavy Breeds are generally intended to be table fowl or to be big-boned, dual-purpose breeds.

HYBRIDS *page 90*

The result of cross-breeding to produce a chicken that is not only healthy and robust, but can produce a large number of eggs. Hybrids are the mainstay of free-range chicken farms across the world.

TRUE BANTAMS *page 80*

These are small, ornamental breeds that have no direct counterpart in the Large Fowl category. They are quite distinct from bantam examples of Light or Heavy Breeds, which are miniature versions of the standard breeds.

LIGHT BREEDS

The crest bears a resemblance to a Swiss lace bonnet.

Below: Black tips to the tail feathers are a distinctive feature of the Appenzeller. They are quite active birds and so are not recommended for first-time keepers.

Appenzeller hens come into lay at about five months and provide plentiful small, white eggs.

Appenzeller

The Appenzeller originates from Switzerland. True to their origins, they are hardy in cold weather. They are active and alert but quite flighty, disposed to roosting in trees. They have a well-spread tail, full hackle, and fairly hard, tight plumage. This striking breed has a forward-facing crest resembling the bonnets worn by Swiss ladies in the Appenzellerland region, a horn-shaped comb, and large nostrils. The breed is found in three colors: silver spangled, gold spangled, and black. The wattles are long and fine and the ear lobes are white and oval-shaped. They have a powerful beak with a fleshy knob at the front. The prominent eyes are brown.

Right: The nostrils are large, the earlobes white, and the eyes alert.

Araucana

Originating in Chile, the Araucana is named after the Arauca Indians. The birds were introduced to Europe in the early 1900s, although they had been known since the mid-16th century. The Araucana has a number of forms. In the United States the rumpless, tailless version with unique ear-tufts that grow from a pad beside the ear is accepted as the true breed. The British developed a tailed version that became standardized in 1969. A tailed version known as the Ameraucana was accepted as a standard breed in the USA in 1984. Famously, Araucanas lay beautiful blue-green eggs. It is possible that the blue-egg gene may be derived from a wild South American fowl called the Chachalaca. The Araucana comes in a range of colors; the self-blue coloring called lavender is the most popular.

Above: A male black-red British Araucana, showing a crest and muffling around the face.

Top right: The eggs of the Araucana are an attractive blue-green color. Here, one is seen with a white and brown egg for comparison.

HANDY HINTS

GO AS THEY PLEASE
Appenzellers are really quite bright and will happily look after themselves when insects are plentiful. They like having freedom to roam as well as having things to climb, as they are always on the go. Appenzellers will need extremely secure fencing as they are very good flyers. They are good layers and will occasionally sit if undisturbed.

FRESH FIELDS
Araucana chicks are strong, fast growers and mature quickly. They tend toward broodiness and make excellent mothers. They do not mind being kept in a pen but like fresh grass so the coop or ark will need to be moved regularly. They are placid birds and are vigorous and hardy. The blue-green eggs are reportedly lower in cholesterol than other eggs.

KEEP IT CLEAN
The tailed version of the Araucana and Ameraucana are relatively easy to look after, even by novice keepers. But you need to keep a check for mites in the crest and muffling. The rumpless, ear-tufted versions are more challenging and so are not ideal for first-time owners.

Light Breeds

Fayoumi

The Fayoumi is an ancient breed and has been raised along the River Nile in Egypt for centuries. They were bred for egg production and are incredibly rugged; reputedly they are highly resistant to bacteria and viruses. They are quite small birds—an adult male only weighs approximately 3–4 lbs (1–2 kg)—with tails held high and large dark eyes. They have a single, medium comb, red earlobes, and slate blue legs. The male has a silver neck and saddle hackles with silver and black barring all over the body. Females also have silver neck feathering and a very darkly barred tail. Fayoumis do not like being handled and will become very vocal when picked up. They can be tamed to a degree with treats but will never be fully tame, and so are not ideal for a novice keeper.

Below: A male Fayoumi. The black and white barring on the plumage is a characteristic feature.

True to their Egyptian origins, Fayoumi chickens cope well with hot conditions.

Right: The Fayoumi matures early and pullets sometimes lay as early as 18 weeks.

Leghorn

This chicken originates from the port of Leghorn in Italy. It arrived in the United States around 1835 when a consignment of brown Leghorns was imported into New York. American breeders recognized its excellent laying ability and by the time the first white Leghorns arrived in the UK in 1870, they were known as superb layers. American and British Leghorns now differ from one another in

Right: A white Leghorn hen of the American type. Leghorns can lay around 280 eggs a year.

appearance. American Leghorns look like the original utility strain with a large tail and medium-sized comb. The UK type now has a shorter tail and very large comb. Both have white earlobes and yellow legs and the eye is red. The Leghorn is active, seldom goes broody, and will provide a good supply of white eggs.

HANDY HINTS

FLYING THE NEST
Fayoumis are lively birds and are very flighty, so care needs to be taken with boundary fencing. Roosting in trees and active foraging is second nature to them, so you will need a pen with a netted roof to avoid them straying too far. They are very good layers of small white eggs with very rich yolks.

QUICK TO GROW
Leghorns are prolific layers— eggs are white and of good size, and are laid throughout the year. Chicks are easy to rear. They feather up quickly, are fast growers, and mature quickly. They can be rather noisy and will roost in trees given the chance. They are not good as table birds as they aren't very meaty.

THE WORLD'S EGG BASKET
The town of Petaluma, California based a vast industry around the commercial egg-laying properties of the Leghorn. In 1916 it held its first Annual Egg Day to celebrate, and it still hosts a Butter and Egg Day Parade in April of each year to celebrate its heritage as the "Egg Basket to the World" during the early 1900s.

Light Breeds

Below: The Lakenvelder color is sometimes described as "a shadow on a sheet."

The head bears a medium-sized comb and the earlobes are white. The black and white belted markings give these birds a very distinctive appearance.

Below: It is unusual to find a neck hackle that does not show some white just behind the head.

Lakenvelder

The origins of the Lakenvelder are not entirely clear. The Dutch claim that the breed takes its name from the village of Lakervelt in the Utrecht region. Germans, however, assert that the breed was developed in the Westphalia area of Germany in the 18th century. The two countries now have different standards for the breed: the Dutch version has a blacker saddle and the undercolor is also darker. Birds should have a medium-sized single comb, white almond-shaped earlobes, and an orange-red eye. The legs are featherless and slate blue and they have four toes. They are a slightly built breed with an elongated body and a tail that is carried high. They make a very good utility bird and have white skin and a particularly plump breast.

Character and temperament

Lakenvelders can be quite high-strung birds. They are light sleepers and will crow if woken by a disturbance in the night; they are often the first breed to sound the alert if any danger threatens. They are not a particularly tame breed, although regular handling will help to calm them down and get them used to your attention. Lakenvelders are a non-sitting breed and rarely, if ever, go broody. They are reasonable layers of medium-sized white eggs, with the bantams often laying more eggs than their larger counterparts. They are very good natural foragers, but not notably destructive of a planted enclosure. They are excellent flyers, able to take off with a 10-foot (3-m) vertical lift from a standstill and can glide more than 33 feet (10 m) from a fencepost. This needs to be borne in mind when planning housing.

Right: A blue Lakenvelder male. Crossing the blue and standard colored Lakenvelder will produce a proportion of chicks of each color.

HANDY HINTS

COLOR MARKINGS
The Lakenvelder is generally seen in the black and white form known as belted. A blue variant in the bantams, in which the black part of the feathering is replaced by an even pigeon blue, was developed in the Netherlands a few years ago. The markings are sometimes rather poetically described as "a shadow on a sheet."

GROWING UP SLOWLY
Chicks mature quickly and grow vigorously, but they don't gain their characteristic markings until they have been through their third molt. Adult males weigh around 4–6 lbs (2.2–2.7 kg) while the females range from 3–4 lbs (1.4–2 kg).

HARD TO BREED
Lakenvelders are not easy birds to breed for showing. A lot of careful selection is needed to obtain the correct color pattern, and it is impossible to tell at an early age which chicks will mature into the best-colored birds, so they have to be kept until they have nearly developed their adult feathering.

Light Breeds

Poland

This is the most popular of the crested breeds and it is also one of the oldest. It has been known as a pure breed since the 16th century. The Poland is a good laying breed, valued for centuries as a producer of large eggs. Polands must be kept dry at all times, so covered runs are essential. The eye-catching crest should be circular and large. It looks amazing, but it does pose problems. It is prone to harboring mites and, if it gets dirty, the bird is unable to preen itself so the job of cleaning falls to the owner. Polands should not be kept with other breeds as they are liable to be bullied and pecked; they are also quite nervous birds. All in all, the Poland is a high-maintenance breed that benefits from the care of an experienced keeper.

Above, top: The laced colors (here a chamois) have beards; non-bearded varieties are also available.

Left: The crest of the Poland (here a silver-laced) should be globe-like. You need to keep these birds out of the rain as a wet crest is not good for them.

Silkie

The Silkie is a very ancient breed; references to it in Chinese texts date back around 2,000 years. They were introduced to Europe in the 1880s. The Silkie is a lightweight chicken with a broad, stout-looking body, that is covered in fine, fluffy feathers that lack barbs. The feathers are not waterproof, so birds must be kept out of bad weather. They have short, rather ragged looking tails and the head is short and neat with an upright and full crest. The Silkie is often crossed with other breeds to produce excellent broodies as they make wonderful mothers. It is a docile, gentle breed, well-suited to a novice keeper who is prepared to give it that extra bit of attention that it needs.

Below: This is a male bantam Silkie. The face of a Silkie should always be dark in color.

WHAT'S IN A NAME?
Surprisingly "Poland" does not indicate that the breed is Polish in origin. The name is thought to derive from the word "polled," which was used to describe the dome-shaped head of some breeds of cattle.

BEHIND YOU!
As a Poland's field of vision is restricted by its large crest, it can easily be startled if you approach it quietly from behind. To avoid unwanted shocks, always speak quietly before picking up a bird so that it knows that you are there.

EYE-CATCHING
Silkies have black skin and dark flesh, turquoise earlobes, feathered legs, and a cushion comb. These features, combined with the fluffy, hair-like appearance of the feathers, make for an extraordinary bird.

Below: A large blue-bearded Silkie pullet.

Light Breeds

Sultan

All present-day Sultans are derived from a small flock of dilapidated birds that were introduced into the UK from Constantinople (Istanbul) in 1854. It is fortunate that they survived, for this is one of the most ornamental birds in the hobby. Sultans have a crest hiding a horn comb, five toes, and large nostrils. The feathering on the legs includes their "vulture hocks"—the quill feathers growing there are stiff and point backward. Sultans are best kept in small runs as they are rather poor foragers. They are calm birds that are not known for aggression and can be handled easily, so they make excellent pets. Sultans are neither table birds nor good layers, but should be enjoyed for their elegance and exceptional beauty.

Always provide Sultans with water fountains to drink from and not open bowls.

Below: The lustrous sheen on the feathers of the black variety of the Sumatra should be green and not purple.

Left: Keeping this plumage clean and tidy is clearly a time-consuming task.

Sumatra

The origins of this breed lie among some fighting fowl that were imported in the USA from Sumatra in the 1840s. The breed was developed away from its fighting roots as an exhibition fowl and its temperament is now as docile as any other soft-feather breed. It is found in black, blue, and white color forms. Both male and female birds have long tails and so special care should be taken to keep the birds clean and their tail feathers unbroken. High perches should be set away from walls to allow extra tail room. The Sumatra lays white eggs and the hens make excellent broodies and mothers. While demanding a bit of extra care and attention, this breed is suitable for novice keepers.

HANDY HINTS

HEALTH CHECKS

Sultans are prone to mite infestation in the crest and beard, and scaly leg mites—all feather-footed breeds are vulnerable to this, so check them regularly. Muddy and wet ground does not suit them as their foot feathers get quickly clogged with mud.

FLYING HIGH

Both Sultans and Sumatras are good flyers and can manage heights of 6 feet quite easily, so care needs to be taken with fences around runs. They need to be tall enough to prevent birds from escaping.

COLORS

Sumatras are found in black, blue, and white forms. Bantam forms also appear in these colors. The black version has a brilliant metallic-green sheen to its feathers.

Above: The Sumatra is a fowl that likes living on an extensive range. They will often take to sleeping in trees, as they have strong wings and fly well.

Light Breeds

Welsummer

Some chickens are bred for their feathers, some for their size, and some for the color and profusion of their eggs. The Welsummer was created by crossbreeding many different varieties of chicken to produce a great all-rounder. The breed was created in the early 20th century and it takes its name from the village of Welsum in the Netherlands. It is thought that Faverolles, Wyandottes, Brahmas, Cochins, Malays, and Dorkings were part of the original mix. Later Barnevelders were added to help create a more consistent shape, and Rhode Island Reds and Partridge Leghorns were also introduced into the equation to boost egg production.

Appearance and character

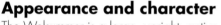

The Welsummer is a large, upright, active bird with a broad back, full breast, and large, full tail. The head has a single comb, medium wattles, almond-shaped earlobes, and a strong, short beak. They have yellow legs, which fade to pale yellow in summer, and reddish bay eyes.

Right: Welsummers are a hardy and attractive breed that generally thrive under free-range conditions. Typically, they display a black-red pattern while the breast is a rich chestnut red. There is also a bantam form.

Welsummers are generally docile birds with a nice disposition.

Below: Welsummer hen. The eggs are a rich, dark brown. They are often exhibited in egg classes at shows.

MOTHERS AND CHICKS

Welsummers do go broody, but not usually until late Spring. However, they are not particularly good mothers. Chicks are strong and are easily sexed, as females have much darker head and back markings than males.

FEATHER-PICKING

Some Welsummer chicks tend to feather-pick if they get bored, so do make sure that the chicks are housed in surroundings that provide them with plenty of mental stimulation to counteract any potential problems.

Below: The hackle feathers of the female Welsummer have a glorious iridescent sheen and reveal a striping pattern that is missing from the males.

Welsummers are always in demand because they lay superb terracotta-colored eggs (sometimes described as flower pots)—up to 200 per year, which can have a variety of speckling and sometimes even look textured. The birds also have attractively colored plumage—males are black and red while hens are partridge-colored. This is an active and alert breed that is a good forager. Welsummers are hardy and generally thrive under free range conditions.

Above: The Welsummer is a very good choice for the novice poultry keeper and is widely kept, so there should be no problems locating stock to buy.

Light Breeds

Below: Vorwerk male. The undercolor of the feathers should be grey and not buff.

The Vorwerk was nearly lost after the Second World War but fortunately was revived from remaining stock.

Vorwerk

The Vorwerk originated in Hamburg, Germany around 1900 and was developed by the German breeder Oskar Vorwerk, who gave his name to the breed. He developed them as good utility birds that would mature early and carry more flesh than many other farmyard breeds. They are powerful and compact birds with broad backs, and a deep, rounded breast.

Below left: Vorwerk blue bantam female. The bantam form appeared in 1997 and the blue variety followed shortly after.

The head is broad and the face well-feathered. They have a single, medium-sized, serrated comb, medium length wattles, and small white earlobes. Their plumage is soft and glossy and is reminiscent of a Lakenvelder with the white replaced by a dark buff. The legs are slate blue and they have four toes. There is also a bantam version. The Vorwerk is a very practical breed that is suitable for a novice keeper. These birds are good layers and mature quite quickly.

Yokohama Red-saddled White

The striking coloration is unique to this breed, which was developed by Hugo de Roi around the area of Brunswick in Germany in the late 19th century. The neck and tail are pure white, while the body is red-buff in the female and a rich crimson color in the male. There is distinct white spangling at the ends of the feathers. The comb is walnut-shaped. The bantam form, which was not standardized until  1968, has now become the most popular type—Large Fowl examples are only rarely seen. Temperamentally, these are equable birds that respond well to handling. They fly extremely well and, given a chance, would prefer to spend the night roosting in a tree than confined to a henhouse!

HANDY HINTS

ONAGADORI

All European long-tailed breeds, like the Yokohama, are related to a Japanese breed called the Onagadori. It contains a non-molting gene and so its feathers grow all year round.

LONG TAILS

The perches for Yokohamas need to be tall and set at a distance from the walls of the henhouse to prevent the long tails from becoming soiled or even breaking. Mud, of course, spells disaster if you are trying to keep your birds looking good.

A deep crimson red color is set against the white background with white spangling on the breast.

Right: A Red-saddled White Yokohama male. This is a rare Large Fowl example.

Top left: A walnut-shaped comb gives this striking breed a very game-like appearance.

HEAVY BREEDS

Australorp

The Australorp was developed as a utility breed in Australia in the 1920s from the black Orpington that had been exported there in the 1880s. They were bred primarily for egg production and they are excellent layers of tinted to brown eggs, and hold the world record for egg production—one Australorp hen laid 364 eggs in 365 days! They are smaller and neater than Orpingtons but still have a lot of meat on them. This is a very practical breed—the birds make good broodies and mothers and are decent layers. The Australorp is active and fast-growing, with hens reaching point of lay at around five months of age. They make good pets as they are calm and friendly birds, although they are a little heavy for children to pick up. Don't let the fact that they are black and so less eye-catching than other breeds put you off the Australorp. It does have a great deal going for it!

Australorps are very hardy and will withstand cold weather well.

Australorps aren't great flyers, so fencing doesn't need to be particularly high to keep them secure.

Right: Australorps have alert, very dark eyes and a wonderful green sheen on their black plumage. Their laying ability makes them an excellent choice for a novice keeper.

arnevelder

he Barnevelder was named after the Dutch town of
Barneveld, which lies about 31 miles (50 km) southwest
of the capital, Amsterdam. This was the center of egg
production in the Netherlands in the early years of the
20th century. A particularly well-balanced chicken, the
Barnevelder has bright orange eyes and clean, yellow legs
and beak. The most popular color is the double-laced,
which has two rings of glossy, greenish-black lacing on a
nut brown background. The Barnevelder is a very solid and
reliable breed that is well-suited to the novice keeper. Hens
lay around 200 caramel brown eggs per year.

Below: A Barnevelder double-laced female. The plumage
is very attractive, especially when well-laced. The best
Barnevelders can easily walk off with top
awards at poultry shows.

The Barnevelder
makes a good,
practical breed for
the novice.

Heavy Breeds

Brahma

These gentle giants of the poultry world make excellent pets for children. Although they are very large—males weigh around 11 lbs (5 kg) and hens around 10 lbs (4.5 kg)—they are placid, good-natured, and easy-going. Brahmas arrived in the UK from the United States in the 1850s. The name Brahma comes from the River Brahmaputra in India, although the species was created in America from large, feather-legged birds known as Shanghais, which were imported from China in the 1840s. These were crossed with Grey Chittagongs from India, which produced the pea comb and beetle brow that we see in the Brahma today. Brahmas are slow-growing, and thus not really suitable as a meat breed. The eggs are surprisingly small and low in number, but chicks hatch strongly and grow quickly. They are trusting birds and are easy to tame. Because of their size, they do take up a lot of space, but they do not fly and can be let out to roam happily in the garden.

Below: The Brahma is a wonderful breed and can be highly recommended to anyone wanting a stunning, placid, and gentle bird to admire.

The blue partridge coloration is a recent addition to the breed.

Note the intricate lacing on this dark female.

Buckeye

Predominantly known as an American breed, the Buckeye was produced by Mrs. Nettie Metcalf in Warren, Ohio in 1896. She crossed Asiatics, Black-Red, and Indian Game fowl. The name Buckeye refers to the color of the plumage, which resembles the buckeye nut found in Ohio. Mrs. Metcalf wanted the plumage to show some dark pigment in the underfluff, and so this part of the feathering is red, except on the back, which should show a bar of slate color when the feathers are parted. The Buckeye is the only purely American breed to sport a pea comb, and this, combined with its stocky build, makes it a supremely cold-hardy chicken. The Buckeye is a tough, robust, dual-purpose breed, but it is quite rare, so locating available stock may take time.

HANDY HINTS

HOUSING TIPS
Perches for Brahmas should be set low to the ground to avoid damage to their legs and feet when they jump down in the morning. Doors must be tall and wide enough to accommodate their size.

KEEP WATCH
Some male Buckeyes have been known to become aggressive during the breeding season, so it is sensible to keep them away from children until you have had time to assess their character and temperament.

Right: Parting the feathers on the back will show a bar of slate coloring in the underfluff. A Buckeye should look very powerful.

Mahogany red is the primary body color while the tail is black.

Heavy Breeds

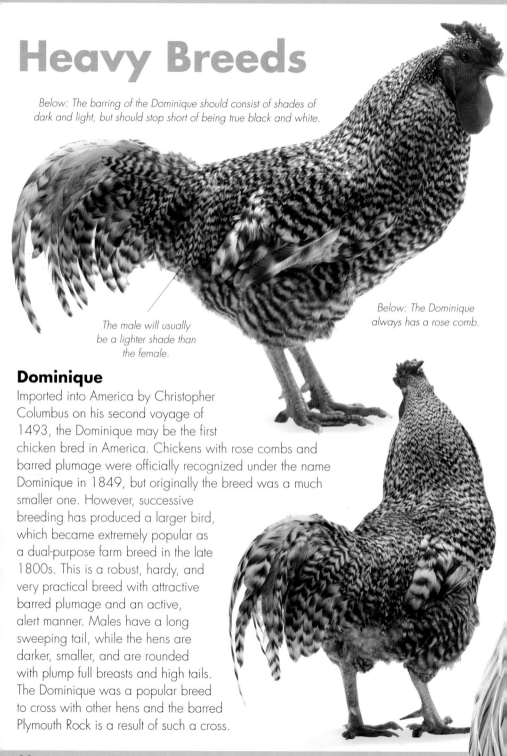

Below: The barring of the Dominique should consist of shades of dark and light, but should stop short of being true black and white.

The male will usually be a lighter shade than the female.

Below: The Dominique always has a rose comb.

Dominique

Imported into America by Christopher Columbus on his second voyage of 1493, the Dominique may be the first chicken bred in America. Chickens with rose combs and barred plumage were officially recognized under the name Dominique in 1849, but originally the breed was a much smaller one. However, successive breeding has produced a larger bird, which became extremely popular as a dual-purpose farm breed in the late 1800s. This is a robust, hardy, and very practical breed with attractive barred plumage and an active, alert manner. Males have a long sweeping tail, while the hens are darker, smaller, and are rounded with plump full breasts and high tails. The Dominique was a popular breed to cross with other hens and the barred Plymouth Rock is a result of such a cross.

Dorking

The Dorking may have arrived in Britain with the invasion of Julius Caesar in 55 B.C. It is recorded that five-toed breeds were lauded by the Romans as the finest type of table fowl, and the Roman invaders probably brought this variety of chicken with them. Whatever its origins, the Dorking has been praised for centuries as the best of British table birds. Dorkings are hardy and quiet in nature. They require space as they are very active and enjoy foraging. Egg-laying tends to take place in the early part of the year. The eggs are of a good size and are white. The Dorking can take up to two years to mature and may live for up to seven years. To keep them at a decent weight and size, they do need to be given a good quality feed. They can become very broody.

Above: A male dark Dorking.

Left: This cock bird displays a large rose comb only found in certain colors.

Right: The red Dorking, here a female, is one of the rarer colors. It always has a single comb.

Heavy Breeds

Jersey Giant

Originally called the Jersey Black Giant after the brothers, John and Thomas Black, who developed the breed, the Jersey Giant is the world's largest chicken. It was developed i the state of New Jersey in the United States between 1870 and 1890 as a dual-purpose bird. The Java was one of the breeds used to create the Jersey Giant; others included the Langshan, Brahma, and Indian Game. The idea was to produce a breed of poultry large and heavy enough to replace the turkey as a premium table bird. Breeders of the time record show birds weighing over 15 lbs (7 kg). Today's weights are more modest, but adult males should still weigh around 13 lbs (6 kg). These are huge birds, well-built with a long, deep breast, with a build described as brick-shaped.

Below: Being strong and hardy, the Jersey Giant is suitable for the novice, although stock is often in short supply.

Colors and character

As well as the original black, Jersey Giants are now found in both white and blue-laced coloration. The legs should be willow (a greenish shade) in color. Adult black birds should develop a green sheen to their feathers.

Below: Jersey Giant chicks. It will take many months and plenty of food before they attain their full size and weight.

Below: The Jersey Giant should have a full, broad, and deep breast, carried well forward.

HANDY HINTS

BREED STANDARD
It took some years for the Jersey Giant to be accepted into the American Standards by the American Poultry Association, but this was finally achieved in 1922.

HOUSING NEEDS
Because of their size, you will need to provide large pop holes and lower perches when housing Jersey Giants. And be prepared for aching arm muscles if you need to carry them any distance.

COLOR VARIETIES
The original Jersey Giants were black in color. The white form was developed shortly after the Second World War, while the beautiful blue-laced variety is a more recent introduction.

A Jersey Giant hen. The undercolor of the feathers of the Black should be slate or grey in color.

The breed is docile and generally well-mannered, and they are hardy in cold weather. These large birds need a lot of space, but are unable to fly, so boundaries do not have to be particularly high to keep them confined. They don't really make ideal pets for small children, however, because of their great size. Hens are reasonable layers and are known for producing eggs during the winter months. They lay around 160 large brown to cream eggs each year.

Left: The eye color in all varieties should be brown or black.

69

Heavy Breeds

Marans

This very popular breed was created through a combination of the Langshan and existing breeds in the Marans area of western France. It started to take on a standardized form in France in the 1920s and was imported into England in 1929, where again it was standardized. The bantam Marans was developed in the 1950s. Marans are a good, robust, heavyweight breed. They are docile birds, quite content in their role as relaxed garden hens, and would make a good choice for a novice keeper. They become very tame if handled frequently and will usually come running across the garden if they suspect you are bringing some tidbits from the kitchen. Marans are consistent layers of large chestnut-colored eggs; you can expect around 200 per year.

Left: This British Marans Large Fowl displays a utility shape. The show birds tend to be a bit heavier.

This hen is cuckoo in color. This means that the barring is soft and indistinct.

Right: A French copper black Marans hen, showing the feathered legs that the French variety still maintains.

New Hampshire Red

This breed was created in the United States entirely from strains of the Rhode Island Red that was already popular. It was designed to improve certain existing traits in the Rhode Island Red, and over the years the color has evolved to a much lighter and brighter red than its ancestors. The head is deep and rather flat on top with prominent eyes, a single comb with five points, smooth face, large wattles, and oval, red earlobes. This a good breed for beginners. They are confident and friendly, lay well (eggs are brown), and are healthy and strong, and have robust chicks. They thrive in a run or wandering free about the garden and, as they are not good flyers, they do not need especially high fencing. They are not aggressive toward each other and are generally tolerant by nature.

HANDY HINTS

MARANS BREEDING
If looking for a cockerel to breed from, buy a clutch of really dark eggs from a breeder. This ensures that any cockerel that hatches will be from a dark egg strain.

RED BANTAMS
New Hampshire Red bantams were developed in the 1940s. As well as Rhode Island Red, there is also Brahma and Wyandotte in their makeup.

Left: New Hampshire Red bantams are fast proving themselves as a practical utility breed of bantam, suitable for back gardens.

Heavy Breeds

Orpington

The early years of the 20th century saw a bout of Orpington hen fever break out as prices for this hugely popular breed rocketed. At a New York show, the sum of $2,500 was paid for a winning Orpington! The breed had been developed in 1886 by William Cook of Orpington, Kent. The Langshan, Plymouth Rock, and Minorca were used in its makeup. The first black form was followed in 1889 by the white, and the well-known buff Orpington arrived in 1894. Within a few years another breeder, Joseph Partington, introduced changes to the breed including a wealth of feathering. William Cook was not happy about this, yet the public loved the fluffy plumage and that is the standard that has persisted.

Right: Buff Orpington male. The Orpington is a large, placid breed.

Above: Lavender Orpington female. The lavender color is not yet standardized, but it is already becoming very popular.

Right: Spangled Orpington female. In other breeds, this attractive black-and-white color is termed "mottled."

Character and size

Orpingtons today are large placid birds with a wealth of feathering. They provide an ideal introduction to the joys of poultry keeping. Adults should typically weigh around 10 lbs (4.5 kg), while the blue variety is even heavier. A bantam form is available if you think that your space is too limited for the Large Fowl type. These birds make good pets as they are docile and not inclined to fly far as they only have small wings and short legs. They like dry conditions and plenty of grass, but have a tendency to overindulge and put on weight—they need plenty of exercise

HANDY HINTS

COLORS
The most widely seen colors are black, buff, white, and blue. Other colors have been developed, although not all are yet standardized. They include gold-laced, silver-laced, lavender, and chocolate.

DUAL-PURPOSE
Orpingtons lay about 175 to 200 medium to large light-brown eggs a year. They do not stop laying in the winter. The chickens also get large, so they are suited for eating.

AUSTRALORPS
William Cook's original black Orpingtons were not lost to history. They were refined as a practical breed in Australia and returned to Britain as the Australorp (see page 62).

Right: Buff Orpington hen. The buff variety carries less fluff around the thigh area than the other colors.

in free-range conditions to keep fit. They have a very strong tendency to go broody and therefore make great mothers. The eggs are small and pinkish.

Heavy Breeds

Plymouth Rock

The name Plymouth Rock was first given to a breed of poultry developed in the late 1840s by Dr. John Bennett of Plymouth, Massachusetts. It turned out to be something of an unpredictable mongrel breed and the strain died out, but the name was revived in 1869 by D.A. Upham who exhibited three birds as Improved Plymouth Rocks. These examples made a great impression and, unlike Bennett, Upham soon had his Rocks

Below: White Plymouth Rock Large Fowl female. The Large Fowl are not as popular as the bantams.

Left: Buff Plymouth Rock bantam. The shade of buff is important; it has to be an even gold-buff right down to the skin.

breeding true to type. Other breeders were working on similarly colored fowl at this time and their efforts eventually all came together under the name Barred Plymouth Rock. They owed their color to the Dominique (see page 66). After the barred variety came buffs, whites, blacks, partridge, blue, and many other colors.

Appearance and character

These are large, long-lived chickens and the hens have a deep, full abdomen, which is a sign of a good layer. They have a broad,

eep, and well-rounded breast and bright yellow legs. The
ıce is red with red earlobes, a bright yellow beak, bay-
olored eyes, and a single medium-sized comb. They are
ɪendly birds that are easy to tame and are vigorous and
ardy. They don't need a lot of space but do appreciate
ıe chance to run free. They are not good flyers so do not
ɪquire high fencing. They lay a reasonable number of
ırge cream-colored eggs, averaging around 200 per
ear, although this can be lower in some strains. Plymouth
Rocks do tend toward broodiness
so regular egg collecting is
important to avoid this.

SIZE MATTERS
Plymouth Rocks are fairly large birds.

A typical male rooster weighs slightly under 10 lbs (4.5 kg) while a hen will be in the region of 7 lbs (3 kg) to 8 lbs (3.5 kg). They are popular dual-purpose birds, valued both for their meat and their egg-laying ability.

POPULARITY STATESIDE
Up until the Second World War, no breed was kept and bred as extensively in the United States as the barred Plymouth Rock. Its popularity sprang from its qualities as an outstanding farm chicken: hardiness, docility, broodiness, and excellent production of both eggs and meat.

Above: White Plymouth Rock Large Fowl male. The legs should be bright yellow. To keep the legs the right color, Plymouth Rocks need access to plenty of grass.

Heavy Breeds

These easy-going birds are well-suited to a garden.

Rhode Island Red

The Rhode Island Red is the famous brown-egg-laying hen seen in farmyards all over the world. It originated in Little Compton in the American state of Rhode Island in the middle of the 19th century. It is a strong and vigorous breed that is capable of laying 260 eggs per year. Because of its productivity and robust nature, the Rhode Island Red was used as the basis for creating many of the hybrid or crossbreed hens popular today in commercial egg-producing farms. They do make good pets as well, but are happiest with plenty of space to range. In 1954 this chicken was adopted as the official state bird of Rhode Island.

Left: The color of the underfluff in a Rhode Island Red should be red or salmon-colored, it should never be pale or white. The legs should be yellow. This is a bantam male.

Left: The deep, broad shape of this bantam female is what the breed standard specifies.

Sussex

The Light Sussex often takes the top honors at shows, and with her pretty black lacing and polar white body leading to a final flourish of black tail feathers, it's easy to see why. Of the pure-breed hens, only the Australorp and Rhode Island Red can rival the Sussex for the number of eggs she lays, up to 260 per year. The eggs are cream to light brown-colored. The docile Sussex hen (there are other colors such as buff, speckled, silver, red, and brown) is quite happy being kept in a garden. There is also a bantam version that lays smaller eggs.

Below: A buff Sussex female. In 1900 the breed was close to dying out, but the Sussex Poultry Club was formed in 1903 and through its efforts the breed was revived and standardized.

HANDY HINTS

DIFFERENT COMBS
Although most often found in the single-combed type, the Rhode Island Red is occasionally seen in a rose-combed variety. A little-known Rhode Island White was also developed between 1918–20 and it formed the basis for some commercial hybrid strains.

BIG BIRDS
A Rhode Island Red is a big bird. A hen should weigh around 5.5 lbs (2.5 kg) and the male 8 lbs (3.6 kg) or more. The body shape is deep, broad, and long, giving the birds an oblong appearance described as brick-shaped.

Above: A light Sussex cockerel. The black and white coloring, known as the Columbian pattern, is the most popular.

Heavy Breeds

Wyandotte

This American breed is part of a very large family. Numerous colors have been standardized and a bantam form is readily available. Originally, it came about because American breeders wanted to create a large, practical fowl that was as pretty as the little Sebright (see page 88) with its fancy lacing. The silver-laced Wyandotte was slowly developed between 1864 and 1872. The name Wyandotte derives from the term that Huron Native Americans used for themselves. A gold-laced variety followed soon after, and then other laced varieties. Then, penciled forms became popular, echoing the delicate penciling found on the Cochin and dark Brahmas.

Right: The flecking on this partridge bantam male's breast shows he was produced from a pullet breeding strain. Males from a cockerel breeding strain have a solid black breast.

The partridge coloration reflects the natural wild pattern that is found in jungle fowl. Males have a black breast and a red or orange neck hackle.

Left: Wyandotte black pullet. The comb of the Wyandotte always fits closely to the head.

Appearance

These are large chickens with a particularly rounded appearance. They have broad bodies and are full-feathered. The hens have a deep breast and backside, which shows that they are good layers. The Wyandotte has yellow legs with a round, short head, a rose comb, bright red earlobes, and reddish-bay eyes. They are a good dual-purpose breed. Wyandottes are docile birds and the hens are excellent broodies and make good mothers. They lay well and chicks tend to be strong and are quick growers. Their attractive curvy shape, generally friendly disposition, and many attractive color forms make them a good choice for novice chicken keepers and show exhibitors alike.

VITAL STATISTICS
Hens usually lay around 200 eggs a year, with an exceptional bird laying around 240 eggs a year. The eggs are brown or tinted. Females weigh around 6 lbs (2.7 kg) and males weigh about 9 lbs (4 kg).

COLORS
In the United States, eight colors are recognized by the APA (American Poultry Association): gold-laced, silver-laced, white, black, buff, Columbian (white with black markings), partridge, and silver penciled. Other national poultry organizations recognize more colors—in the UK 14 are standardized.

Right: The lavender color seen here is a recent creation. Wyandottes are available in a wonderful variety of colors.

TRUE BANTAMS

Barbu d'Anvers

This bantam—a true dandy of the poultry world—is Belgian in origin. Its popularity in the UK took off in the early 20th century when birds were exhibited at the Crystal Palace Show. The breed should be as small as possible; birds have a rose comb and should show abundant beard and muffling. The face should look owl-like. The most common color is quail, a combination of golden buff and black. Females are sweet-natured and tame easily. Males are supreme egotists, strutting and posing around the yard. This makes them excellent show birds. Despite their small size, these birds are hardy and do well on an extensive range.

Left: Barbu d'Anvers lavender quail male. In the breeding season the males will be protective of their hens, but females are always sweet-natured.

Left: Barbu d'Anvers blue quail male. The pointed end at the back of the rose comb should follow the line of the neck.

Below: The white areas of the millefleur plumage will increase as this Barbu d'Uccle hen grows older.

HANDY HINTS

FEISTY NATURE
Barbu d'Anvers cockerels have a reputation for occasional aggression toward their owners, so keep them away from small children until you have gauged their individual temperaments.

BUYING ADVICE
Barbu d'Uccle bantams can be prone to Marek's Disease, a viral condition that causes paralysis and even death. Check with the breeder if the stock has been vaccinated before you decide to buy new birds.

COUNTRY COUSINS
The Barbu de Watermael (below) is a variety of the Barbu d'Anvers, which has a crest, as well as the characteristic beard and muffling. Another rumpless variety that lacks a tail is called the Barbu du Grubbe.

arbu d'Uccle

t is thought that this breed was developed in the 1890s rom the Barbu d'Anvers and the Booted Bantams (see page 82). It is also Belgian in origin, coming rom the municipality of Uccle, near Brussels. This bantam should have a wealth of muffling, a single comb, and feathered legs and feet. The attractive millefleur color is a rich mahogany with feathers ending in black with a white tip. Another color form, porcelain, is pale straw with the black in the feathers replaced by lavender. Female birds have a very sweet nature and can become extremely tame. Males, however, like the Barbu d'Anvers, can have an aggressive streak. These birds do well on an extensive range, but do keep them off muddy ground, which would soil the leg feathering.

True Bantams

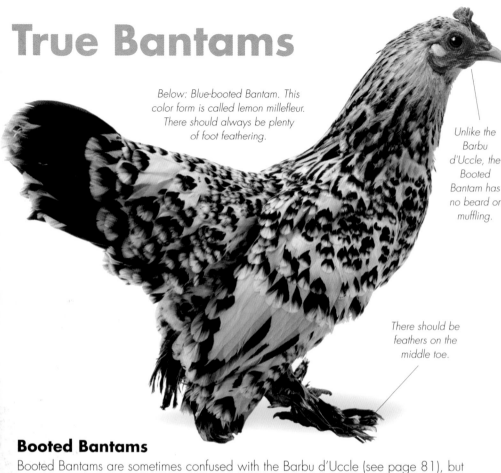

Below: Blue-booted Bantam. This color form is called lemon millefleur. There should always be plenty of foot feathering.

Unlike the Barbu d'Uccle, the Booted Bantam has no beard or muffling.

There should be feathers on the middle toe.

Booted Bantams

Booted Bantams are sometimes confused with the Barbu d'Uccle (see page 81), but the absence of beards or muffling around the face readily distinguishes this breed from the Belgian variety. These sprightly little chickens appear in an attractive range of colors, such as silver millefleur, buff mottled, and lavender mottled. This breed is particularly popular in Europe, and may be Dutch in origin. Records of feather-legged bantams stretch back as far as 1600. They are also known as Sabelpoot, which translates from Dutch as "sword leg," a reference to the stiff feathers that grow on the hocks. Keeping these feathers clean can be problematic, but if the birds do not have to be kept looking pristine for a show, they do very well outdoors.

Dutch

These tiny bantams—males only weigh around 1 lb (500 g) and females are even smaller—were standardized in the Netherlands in 1906. They are upright little birds with short backs and a high, full breast. The wings are fairly large and long and are carried close to the body. Despite its small stature, this is a very robust breed that suffers from few health complaints. It is an ideal choice for anyone wanting to keep a True Bantam for the first time. It comes in a huge range of colors and will be readily seen at poultry shows. Although principally kept for ornamental purposes, they are good layers, good setters, and good broodies. Because of their small size, Dutch females are only capable of covering a small clutch of eggs. Dutch bantams are jaunty little birds but need to be protected from the winter weather. They also need secure fencing as they are good flyers.

Dutch bantams have a low wing carriage, as in this male.

HEALTH CHECKS
Like Barbu d'Uccles, certain strains of Booted Bantams are susceptible to the potentially fatal viral condition called Marek's Disease. It makes sense to check whether your breeder has vaccinated against this.

IN THE DRINK
Dutch chicks should be provided with shallow drinkers as they are liable to drown if they happen to fall into a deep water container.

Left: The Dutch is available in a wonderful range of colors—this is millefleur. The earlobes should be white and oval-shaped.

True Bantams

Japanese

Developed in Japan as early as the 7th century, probably from Indo-Chinese stock, the Japanese or Chabo bantam was first introduced to Europe around the 16th century. It has very short, yellow legs and the long tail is carried high and points forward. This is normally described as being squirrel-tailed. As a result, the body is almost U-shaped, with the wings held so low that the tips touch the ground. Japanese bantams have an evenly serrated single comb that, in the male, tends to be rather large, and the face and ear lobes are bright red. They have a rather waddling gait due to their short legs and broad build. They should be kept in a clean henhouse, as their wingtips touch the ground and can be easily soiled. They may suffer outside in bad weather because of their short legs and fancy feathering, but they are well suited to being kept in confinement in gardens.

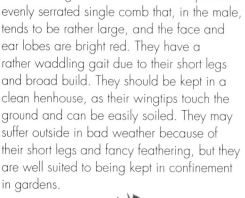

Right: A female Japanese bantam. This is a suitable breed for any novice prepared to provide a bit of extra care. This hen is actually standing up.

Buyers should look for very short legs and an upright tail.

Nankin

The Nankin takes its name from a yellow cotton cloth known as nankeen that was widely used in the latter part of the 18th century to produce waistcoats and breeches. The bird's color closely resembles the color of nankeen. Nankins should appear neat and bold. The breast is carried well forward, the back is sloping, and the wings are carried low, nearly touching the ground. Two types of comb are permitted—the single-combed variety is seen most often, but a rose-combed variety may also be encountered. The legs should be blue or bluish-white. These birds are reasonable layers by True Bantam standards and make good broodies and mothers.

Both above: This bird displays the rich yellow coloring that gave the breed its name.

True Bantams

Pekin

This bantam has quite an exotic story attached to it. During the Opium Wars between Britain and China in the middle of the 19th century, British troops looted the Emperor's palace in Peking (Beijing). It appears that among the spoils shipped back to the United Kingdom were examples of this bird. It quickly became popular in Britain and more birds were imported in the years that followed. There is no evidence that Pekins were ever produced by miniaturizing the larger Cochin breed, although it is sometimes referred to by the name Cochin bantam. It is more likely that the birds were carefully bred as Imperial pets from the strains of feather-legged bantams that were prevalent at that time.

Below: Blue mottled Pekin pullet. This breed has a bright personality, loves being fussed over, and really seems to enjoy human company.

Below: Lemon cuckoo Pekin female.

MOTHERLY LOVE
While Pekins are kept principally for their looks and because they make excellent pets, they do also lay a reasonable number of eggs. They make good broody hens and are attentive and careful mothers.

SMALL SIZE
Shorter than the ordinary bantam, Pekins are often only about 8–12 in (20–30 cm) tall (with head upright). Their carriage tilts forward, with the head slightly closer to the ground than their elaborately feathered tail.

DIFFERENT NAMES
In Europe and the United States, the Pekin bantam is known as the Cochin bantam. The American strain is larger and broader than its UK equivalent.

Appearance and character

A Pekin should remind you of a round ball of feathers, and its cushion-like tail looks like a rounded extension of the rest of the bird. Legs are very short, so much so that you may even be mistaken into thinking that it is sitting down. Pekins are available in a dazzling variety of colors, and there is also a version in which the feathers curl outward, all of them facing toward the head. The feather type is called a Frizzle. Pekins are not destructive when allowed to roam around the garden—their foot feathering prevents them from doing much damage to plants. This feathering does mean that Pekins need more attention than some breeds to keep them looking good, but their docile temperament makes them ideal for novice keepers and they are very good pets for children.

A lavender Pekin male (right) and female (far right).

True Bantams

Sebright

This beautiful bird originated in Britain. At the start of the 19th century a group of poultry fanciers set out to create a perfect fowl. Pre-eminent among them was Sir John Sebright, the MP (Member of Parliament) for Hertfordshire and a well-known livestock breeder. The precise strains that went into the makeup of the Sebright were not recorded, but it seems likely that a Nankin (see page 85) and a pit-game cock were involved. The spectacular lacing in the feathers may have been introduced after a laced bird, possibly a laced Poland, was acquired from a local zoo. Each feather is edged with black, making the ground color appear brighter and more striking. The first color created was gold, and this was followed by silver.

Below: A silver Sebright male. Although the standard calls for a mulberry-colored face, it is seldom seen in the males.

Male Sebrights lack pronounced sickle feathers in their tails. For this reason they are described as being "hen-feathered."

The legs should be free of feathering and slate blue in color. The feet should have four toes and also be slate blue.

Below: A gold Sebright female. Each feather should be evenly and clearly laced with black.

The feathers are almond-shaped.

Star attraction

This is probably one of the most beautiful bantams in the hobby, but it must be noted that it is not for the novice keeper. Sebrights are not easy to keep as they can be somewhat neurotic. Introducing new hens to a breeding pen can be difficult, as the existing hens can take exception to a newcomer and act aggressively toward her. For this reason, it is often best to run this breed as pairs. The hen is not a great layer—perhaps 60–80 creamy white eggs a season—and the eggs themselves are small. However, for more experienced poultry fanciers, these laced gems are worth all the effort. They are an absolute joy to watch as they strut around the garden.

HANDY HINTS

HEN-FEATHERED
Sebright males are hen-feathered. This means that they do not exhibit any sickle feathers (the long curving feathers in the tail of a rooster) and so their tails look like those of the females. Both male and female birds share the beautifully laced plumage all over their bodies.

HEALTH CHECK
Like a number of other bantam breeds, Sebrights are susceptible to the potentially fatal viral condition called Marek's Disease. It makes sense to check whether your breeder has vaccinated against this.

SMALL IN STATURE
The Sebright is one of the smallest bantams available. Males weigh an average of 1.4 lbs (625 g) and females just 1.2 lbs (570 g).

These feathers are clear. You do not want to see dark spots (or smutting) in the feathers.

HYBRID VARIETIES

More eggs

Although the pure breeds can look beautiful with amazing feathers and combs, they tend to sacrifice performance for looks. Crossbreeding two or more pure-breed hens can result in a chicken that not only looks good, but also lays a super number of eggs. These birds are called hybrid or crossbreed chickens and can be found on free-range chicken farms all over the world.

Right: Hybrids have been developed commercially to maximize egg-laying or meat-producing ability. Hybrids are popular with first-time chicken keepers and make affectionate pets.

Over the years, people have created crossbreeds, which make excellent pets for the first-time chicken owner because of their friendly nature and ability to consistently lay well. Breeders will often give the crossbreeds their own particular names, so it is best to ask about the particular characteristics of the chicken that you are interested in.

Rhode Island Red-based hybrids

There are many hybrids to choose from, with most laying in the region of 300 eggs a year. Lots of them are based on the Rhode Island Red, a dark brown, pure-breed chicken that lays around 250 eggs per year. Crossing the Rhode Island Red with other breeds—such as the Light Sussex, Marans, and Plymouth Rock—has produced lots of differently colored hybrids that will lay you an egg almost every day. The cross-breeding also results in differently colored eggs. Crossing with a Marans produces a slightly darker egg than usual—a pale chestnut brown. Spending time with these chickens is soon rewarded as they become amazingly tame and comfortable with human contact.

White Star

Although most of the hybrids are good layers, there are other appealing characteristics to look out for. The White Star (right) is a beautiful-looking chicken with the most amazing big, floppy red comb, due to its Leghorn ancestry. It lays a good number of pure white eggs but takes much longer to get used to human contact. It can be quite skittish, although it can be highly amusing seeing it make cartoon-style sprints across the garden!

HANDY HINTS

PEACE OF MIND
Buying hybrids from a breeder is a good way of ensuring that you have purchased hardy, healthy birds. Vaccination is a relatively expensive procedure, but the larger producers do it as a matter of course.

DO A GOOD DEED
Some people have set up schemes to rescue battery hens. Although they may look worse for wear, these poor chickens still have a lot to give and by buying them you can ensure they live out the rest of their days in relative luxury.

Hybrid Varieties

Black Rock

The Black Rock was produced in America and is one of the most successful crossbred chickens. Bred from specially selected strains of Rhode Island Red cockerels and Plymouth Rock females, the hens are friendly, not easily stressed by being handled, and superb egg layers. In the right conditions, they will lay up to 320 eggs per year. They have thick plumages, don't mind getting a bit wet, and enjoy human company—so much so that they quite often follow their owner through the back door and into the kitchen if given a chance to do so.

Below: This is a cross between a Rhode Island Red and a cuckoo Marans. She is docile and lays brown eggs.

Above: Black Rocks are one of the most successful hybrids for free-range poultry keeping. They are resistant to parasites and generally healthy and hardy.

Right: Poised and elegant, a good all-around chicken.

Speckledy hen

Speckledy hens are very attractive hybrid chickens (Rhode Island Red crossed with Marans) that make excellent garden pets. They have white legs and barred dark grey and white feathers with a good sized comb and wattles. A Speckledy chicken will lay around 250 eggs per year, and because they are bred from the French Marans hen, lay a very dark brown egg. They look very similar to a pure-bred Marans and share quite a lot of its characteristics, such as being active, graceful, and placid. They are reasonably heavy chickens and will eat quite a bit more than other breeds.

HANDY HINTS

Right: The Speckledy, a Marans/Rhode Island Red cross, can produce around 250 dark brown eggs a year.

GLOSSARY

Bantam Technically the "bantam" is a type of fowl that does not have a larger version. There are several of these "True Bantam" breeds. Many small fowl are also referred to as bantams but they are actually "miniatures" or a small version of a Large Fowl breed.

Barring A black and white pattern across the feathers.

Bloodspot An egg defect, caused by the rupture of blood vessels in the chicken. They are unsightly but still edible.

Broody The desire of a hen to sit and hatch eggs.

Chicken Technically the name for a bird (male or female) of the current season's breeding.

Clutch The number of eggs laid by a bird.

Cock A male bird after its first molt.

Cockerel A young male bird.

Comb The red fleshy growth on the head of most chickens.

Crest The bunch of feathers on the head of some breeds.

Crop Part of the pre-digestive system of the chicken. Food collects here at the base of the neck and is softened before passing through the rest of the digestive process.

Cuticle The last layer applied to the egg in the hen's vagina. It acts as a barrier to disease-causing organisms.

Drinker A container for water that birds can drink from.

Dustbath Chickens will use an area of dry dust, be it earth or sand, in which to bathe in order to remove mites and lice from their feathers.

Earlobe The fleshy part by the ears.

Flight feathers The largest feathers on the edge of the outstretched wing.

Gizzard The internal organ of the chicken that collects grit and aids the digestive process by grinding food down.

Grit Insoluble stony matter fed to chickens to assist the gizzard in grinding up their food.

Hackle One of the long, narrow feathers on the neck or saddle of a bird.

Hard Feather A term used to describe game breeds that were originally bred for fighting.

Hen A female after her first laying period, roughly a year and a half old.

Horn comb A comb that resembles the letter "V," often found in breeds with crests.

Hybrids Birds that have been genetically bred from two different breeds to embody good characteristics from each, such as laying well and developing a good amount of meat.

Keel The bird's breast bone.

Lacing A line of different color found around the edge of a feather.

Large Fowl A standard-sized chicken, as opposed to a smaller bantam.

Meal A mixture of (wet or dry) coarse ground feed.

Molt The yearly shedding and replacement of feathers. It lasts for around eight weeks.

Muffling Feathers found around the beak.

Pea comb A comb that looks like three separate combs, the middle one being the largest.

Pellet Type of food formed from a fine mash bonded together into small pieces.

Penciling Either describing small stripes seen in bands across feathers, or multiple concentric lines following the feather's outline.

Point of lay Term applied to a young pullet at about 18 weeks old, the age at which the bird could start laying. The first egg may not appear until four weeks after this, however.

Primary feathers The first ten feathers on the wing starting at the tip and working toward the body. Out of sight when the bird is resting.

Pullet A female bird from the current year's breeding.

Pure breed A breed that is pure, i.e., it has not been crossed with other breeds or different varieties of the same breed.

Rose comb A wide comb that is nearly flat on top, and covered with small nodules (or "workings") ending with a spike. Its size varies with the breed.

Scales The horny skin tissue covering the toes and legs.

Single comb A flat vertical comb with serrations along the edge.

Soft Feather All breeds of poultry that were not bred for fighting purposes.

Vent The orifice at the rear end of the bird through which both eggs and feces are ejected.

Wattles The fleshy appendages hanging down either side of the lower beak.

Wing clipping The practice of clipping (cutting the ends off) the primary and secondary feathers on one wing to prevent the bird from flying.

INDEX

Index